720.952 SUM
ST

New Architecture in Japan

Yuki Sumner
Naomi Pollock
with David Littlefield

Photography by
Edmund Sumner

New Architecture in Japan

MERRELL
LONDON · NEW YORK

First published 2010 by

Merrell Publishers Limited
81 Southwark Street
London SE1 0HX

merrellpublishers.com

Text copyright © 2010 Yuki Sumner, Naomi Pollock and David Littlefield, except page 7, copyright © 2010 Takero Shimazaki

Illustrations copyright © 2010 Edmund Sumner, except pages 84–85, copyright © 2010 Dennis Gilbert, and pages 30t, 76t and b, 114–15, 148tl and 240–41, copyright © the individual architects

Plans and renderings copyright © the individual architects

Design and layout copyright © 2010 Merrell Publishers Limited

All rights reserved. No part of this publication may be reproduced, stored in a retrieval system or transmitted, in any form or by any means, electronic, mechanical, photocopying, recording or otherwise, without the prior written permission of the publisher.

British Library Cataloguing-in-Publication data:

Sumner, Yuki.
New architecture in Japan.
1. Architecture – Japan – History – 21st century.
I. Title II. Pollock, Naomi R. III. Littlefield, David.
IV. Sumner, Edmund.
720.9'52'090511-dc22

ISBN 978-1-8589-4450-0

Produced by Merrell Publishers Limited
Designed by Nicola Bailey
Project-managed by Rosanna Lewis
Indexed by Diana LeCore

Printed and bound in China

Front cover: House N (pages 200–201)
Page 6: Gae House (pages 192–93)
Pages 8–9: Final Wooden House (pages 100–101)
Pages 20–21: Chichu Art Museum (pages 64–65)
Pages 28–29: Yokohama International Port Terminal (pages 54–55)
Pages 56–57: Ron Mueck's *Standing Woman* in Towada Art Center (pages 90–91)
Pages 94–95: Takasugi-an (pages 122–23)
Pages 132–33: Ina-Higashi Elementary School (pages 148–49)
Pages 166–67: White Chapel (pages 182–83)
Pages 184–85: House O (pages 202–203)
Pages 232–33: Tod's Omotesando Building (pages 262–63)
Back cover: Final Wooden House (pages 100–101)

The authors are grateful to the following for their assistance in the preparation of this book:

7 Foreword
Takero Shimazaki

11 The Residue of Japan-ness
Yuki Sumner

23 Architecture in Japan: In Context
Naomi Pollock

28 Infrastructure and Public Spaces

56 Culture

94 Sport and Leisure

132 Education

166 Health and Religion

184 Houses and Housing

232 Offices and Retail

268 Further Reading

269 Index

272 Acknowledgements

272 Biographies

Foreword
Takero Shimazaki

I had the privilege of working with the photographer Edmund Sumner on one of the buildings my practice completed in 2009. His sense of freely zooming in and out was one of the skills that I found refreshing and particularly energetic. He would walk across the field as far as possible and climb up the hill to capture the small elements of the building that appeared between the houses and the landscape in the foreground. Buildings are not stand-alone objects that look great only in photographs and on the covers of magazines. Edmund's tactile approach to architecture and its context seems to acknowledge and reinforce this thinking.

Japanese architecture, when documented for those outside the country, is often presented as a series of beautiful, strange, sculpted and ephemeral abstract objects. It can be very hard to understand the political, social and geographical context until you visit. Edmund prefers to include the surrounding chaos (such as electricity cables), local life and the contradictions of context in his photography to capture the sense of being there.

When Edmund and his partner, Yuki, told me that they were embarking on the trip for this book, travelling from Hokkaido to Kyushu for three months with a toddler, and with another baby on the way, I could not quite grasp the task they had set themselves. When the draft of this book arrived on my desk, I was astounded by the distance they had covered, and also by the diversity of the projects they had visited.

Edmund's skill at immersing himself in the local culture (because of his many visits and his understanding of it), as an outsider, is a great tool with which to document the contemporary Japan. He seems to be able to capture the sense of the architects' works, as well as sometimes simply cutting himself out of the insular world that is the Japanese architectural scene.

Japanese architects now influence the global agenda, yet there is also still a strong sense of isolation from the rest of the world. With the global recession, there appears to be even more focus on 'looking within', with more architects working locally and the 'West' no longer the centre of admiration. It also seems that the rigidity of the hierarchy is diminishing, giving the younger generation of architects more opportunities.

In these challenging times it is very interesting to contemplate how architecture will develop. There is a fascinating mix of architects in Japan: some focusing on locality, history and even the chaos of cities, such as Hiroshi Sambuichi, Terunobu Fujimori and Atelier Bow-Wow; and others capturing local and global agendas simultaneously, such as Ryue Nishizawa, Sou Fujimoto, Kumiko Inui, Tezuka Architects and Junya Ishigami.

Edmund has managed to document the seeds and fruits of these current proposals, with their varying approaches, by freely hopping from one prefecture to another, and by focusing in on and out of the details. This book represents the enormous potential that exists in Japan for the future creation of an enduring and rich architectural heritage.

The Residue of Japan-ness
Yuki Sumner

Roland Barthes once wrote that architecture is the convergence of dream and function.[1] The images created in the 1960s by the Metabolists, of buildings as living organisms capable of multiplying, adapting and changing form, were the very first architectural dreams to emerge from Japan, and captivated Western audiences. Kenzo Tange's Plan for Tokyo (1960), Kisho Kurakawa's Helix City (1961), Kiyonori Kikutake's Ocean City (1962), Arata Isozaki's City in the Air of the same year, Fumihiko Maki's Golgi Structures (1967) and others were only models or drawings on paper, but they conveyed the group's visions with a fervency that led the architectural critic Kenneth Frampton to call the movement 'frantic futurism'.[2] In 1964 the group's teacher, Kenzo Tange, finally proved to the world with his National Gymnasium for the Tokyo Olympic Games that Japan could not just copy but also equal first-rate modern architecture.

Many of the Metabolists' ideas required the remodelling of entire cities, and a clean break from history – unthinkable in such places as Britain and Europe, where the present is heavily weighed down by the past. Reduced to mere dreaming, the British group Archigram 'was more interested in the seductive appeal of space-age imagery and, after [the American architect Buckminster] Fuller, in the Armageddon overtones of survival technology than in the processes of production or the relevance of such sophisticated technique to the tasks of the moment'.[3]

As Japan, its cities flattened, struggled to recover from the devastation of the Second World War, the Metabolists – unlike Archigram – had every intention of building their projects, in order to relieve the very real social problem of overcrowding caused by Japan's rapid urbanization. Some managed to realize their visions, albeit in drastically subdued forms. Kisho Kurokawa's Nagakin Capsule Tower in Tokyo, built in 1972, is a famous example.[4] More recently, Kiyonori Kikutake designed the Global Loop for Aichi Expo 2005 (page 13) – an elevated walkway that connected the six groups of pavilions, allowing, the architect claimed, people of all ages to experience the Expo in one quick sweep. The idea was based on the much wilder scheme that Kikutake had proposed in 1992 for the Earth Summit in Rio de Janeiro, Brazil, whereby the various infrastructures of a man-made city were fitted inside a large doughnut-shaped capsule hovering on stilts high over the Amazonian jungle. In Japan, functionality was never renounced in the face of a powerful dream.

The speed of Japan's conversion to futurism furthermore increased the artificial sense of self, to be moulded and whipped up by external factors. This may be a critical disjuncture by which modern Japanese architecture was consolidated, and by which, to some extent, the country's contemporary architecture is still affected. Arata Isozaki argues eloquently[5] in his most recent book that the identification of Japan-ness began with 'an external gaze': 'For Japanese Modernists – and I include myself – it is impossible not to begin with Western concepts. That is to say, we all begin with a modicum of alienation ….'[6]

Despite the government's attempts at standardization during the post-war period, an exuberantly heterogeneous group of buildings – inevitably both good and bad – emerged as the country modernized itself at incredible speed. The Tokyo Olympic Games in 1964 were decisive in generating the infrastructure on which Tokyo, and many major cities modelled on it, still relies. Instead of building bigger and more efficient roads, for example, the government hastily decided to avoid where possible the negotiations necessary to buy privately owned patches of land, with the result that most of Tokyo's super-expressways are built over existing roads and rivers. In his book *The Making of Modern Japan*, Marius B. Jansen bemoans the fact that 'Nihonbashi, the Japan Bridge, so striking and busy in Edo and Meiji times, now lies hidden beneath the

Takasugi-an by Terunobu Fujimori is the architect's take on the traditional private tea house.

elevated roadways of the contemporary metropolis.'[7] Osaka City followed suit, erecting a vertical city where three or more lanes of traffic constantly diverge and converge high above the ground, much more intensely than in Tokyo.[8]

It is unsurprising, then, that the contemporary architect Tadao Ando, operating as he does from such an extreme city as Osaka, seeks to filter its lawlessness by paring down his structures to concrete and steel, with only narrow (usually horizontal) openings in the façades.[9] Others follow closely in his footsteps: Kei'ichi Irie of Power Unit Studio managed to convince dexterous Japanese builders to mould concrete in anti-gravitational ways for his Y House (pages 228–29), shutting out once and for all the views of neighbouring houses in the suburbs of Nagoya City; and Shuhei Endo's Rooftecture S (pages 220–21) edits views with its cladding of galvanized steel, its robust quality allowing the house to cling to the side of a cliff like a mountain goat in a suit of armour. In this book, however, we shall see that a more recent trend is to embrace visual chaos through loosening of form. A Miesian rectilinear glass box is deconstructed in Sou Fujimoto's House O (pages 202–203) and in Ryue Nishizawa's Towada Art Center (pages 90–91), letting the outside world seep in from various directions.[10]

Tokyo is the city from which modern Japanese architecture has evolved. Kurakawa, one of the post-war generation of architects, described the metropolis as a large meadow, into which a 'wild' rabbit like him went in search of a great life.[11] The city was an extension of his own home: he could, if he wanted, eat out three times a day, meet his friends and enjoy hot baths in *sento* (communal bathhouses). Surprisingly, more than sixty years after the end of the war, the young architect Sou Fujimoto talks with similar enthusiasm about the 'wilderness' of Tokyo, where small alleys wriggle through the jumble of tall buildings, irresistible in its disorderly unpredictability to someone like him, who grew up in central Hokkaido:

'I enjoy the city as I used to enjoy the hills and paddy fields of the countryside in which I grew up.'[12]

Just as the marginality of Fujimoto's background has greatly informed his work, so some architects are resisting the one-way system of cultural assimilation. Shin Egashira and a group of students from London's Architectural Association School of Architecture, for example, make annual expeditions to the small village of Koshirakura in Niigata Prefecture (pages 76–77). Every summer the whole village looks forward to the influx of young foreigners, who are eager to learn local crafts and carpentry from the (mostly elderly) locals. In return, they provide a unique boost to the village's fragile infrastructure by creating experimental buildings for it. The students have also become an indispensable part of the regional festival, during which a gigantic tree trunk is carried down the mountain into the village. Such an intense, albeit temporary, intoxicating cocktail of different cultures – peripheral and central, young and old, past and present, memories and perceptions – leaves unique and defiant landmarks in the remote countryside.

Based in Hiroshima City, and thus also operating outside Japan's architectural centre, another young architect, Hiroshi Sambuichi, is inspired by history. His low-rise Inujima Art Project: Seirensho (opposite; pages 70–73) resembles a large tumulus, preserving – or burying – a cultural past (for example, a house once owned by the writer Yukio Mishima has been demolished, moved and partly resurrected by the artist Yukinori Yanagi), as well as an industrial one.[13] Sambuichi's sensitive approach to recent Japanese history is unusual, and may partly be a result of a life spent in his home town of Hiroshima, where Japan's only ruin, the Hiroshima Genbaku Dome, still stands.[14]

The attitude of the Japanese to history is ambiguous. The legacy of the Second World War, during which much of the country's infrastructure was destroyed, goes hand in hand with that of Japanese imperialism in China, Korea, South-East Asia and the Pacific:

Kiyonori Kikutake's elevated walkway for Aichi Expo 2005.

Hiroshi Sambuichi's subtle intervention on the island of Inujima.

'[The war] was brought about by the very aberration of the modernization itself', Kenzaburo Oe pointed out in his sober acceptance speech for the Nobel Prize in Literature in 1994. Seeking the definition of 'Japan-ness' is therefore a highly problematic exercise, but one of which we Japanese never tire. We must seek its definition either by obliterating the past with a vision of the future (the path the Japanese Modernists inevitably took), or by remaining idealistically vague and inscrutable; we cannot afford to probe too deeply into the past or the murky areas of our own psyche. Oe, baffled by the vagueness of the speech made by his predecessor Yasunari Kawabata in accepting the same prize in 1976, commented that 'even as a twentieth-century writer Kawabata depicts his state of mind in terms of the poems written by medieval Zen monks', the essence of which is 'the linguistic impossibility of telling [the] truth'.[15] Isozaki echoes Oe's concern that somehow the vague 'layer of ancient words' can make the Japanese stop questioning.[16] Japan's ambiguity, or its inability to define itself clearly, is, said Oe, 'a kind of chronic disease that has been prevalent throughout the modern age'.[17]

SANAA, a practice that excels in creating lightweight, transparent, well-programmed spaces, directs its gaze firmly on the here and now: 'It is all about the context .... We do not transform Japanese elements into our own architectural language. We might be inspired by history or tradition but this could come from any country or culture.'[18] Any 'Japan-ness' in their work comes from their refusal to be rooted in it; their lightness of touch and use of flowing space, however, are both undeniably Japanese in character, and can be traced back to the work of an earlier generation of architects, such as Fumihiko Maki and Yoshio Taniguchi. The architect and writer Thomas Daniell, who lives and works in Kyoto, sums up SANAA's work in his recent book *After the Crash: Architecture in Post-Bubble Japan*: 'These buildings seem intended as refuges from their surroundings,

THE RESIDUE OF JAPAN-NESS

Large windows on Moriyama House by Ryue Nishizawa directly look on to the neighbouring houses.

Makoto Yokomizo has created a fluid circulatory system for Tomihiro Art Museum.

moments of provisional silence and stasis within a corrosive context of speed, confusion, and pollution.'[19]

Unlike Ando, SANAA adopts a strategy not to block out the chaos but to be slightly detached from it. Its soft, almost non-existent moulds, with well-positioned mirrors and windows, do not protect us from the exterior gaze, but rather expose us to it, as if to suggest, 'We have nothing to hide.' Toyo Ito was the first to explore this highly revealing form of architecture. The transparent façade of his Sendai Mediatheque (pages 84–85) exposes the diverse activities within to people strolling outside.[20] Seemingly modern and democratic, this strategy also strikes a chord with the Japanese need to reiterate the myth that at the heart of all things Japanese there is a void, a myth that is, of course, not linguistically definable. This strategy can, however (and must, if it is to avoid any hint of phoney Zen anachronism), question the ambiguity of modern Japanese identity. The openness of Ryue Nishizawa's Moriyama House (left, top; pages 210–11) confronts the inwardness of typical post-war Japanese homes.

In this new type of architecture, then, we must become self-conscious beings; we are in charge as the navigators and signifiers of our own surroundings. However, for a nation used to gardens and parks endowed with set narratives (traditional or otherwise – as indicated by the popularity of theme parks in Japan), such an idea is not welcomed wholeheartedly without a great leap of faith. Even the staff of SANAA's 21st Century Museum of Contemporary Art, Kanazawa (pages 60–61), have had to put up partitions inside the museum to stop visitors going in the wrong direction, although such partitions directly contradict the aim of the architect's concept, which is to let people wander freely and aimlessly.[21] Similarly, Makoto Yokomizo's elegant scheme for the Tomihiro Art Museum (left; pages 88–89), its overlapping circles – like ripples made by an insect on the surface of a pond – enabling even wheelchair users to navigate freely in and out of the circular exhibition rooms, is sadly sabotaged by the museum's protocol, with partitions and signs with large arrows.²² Recently this trend (the free-roaming structure characterized by a lack of hard boundaries) has been pushed to an extreme by the younger generation of architects, such as TNA, Junya Ishigami (right) and Kumiko Inui, in whose work people or things become unwitting performers or props. Almost exhibitionistic in character, their buildings are designed with an audience in mind.

If Japan's ambiguity creates a comfort zone within which its traumatic past can be buried, then subtle subversions of the norm offer a temporary escape from the stifling rules and regulations, rituals and protocols that plague its society.²³ In a recent interview, the American poet Gary Snyder remarks that 'Japan's strength is that under all its "strength" is a softness and warm craziness, an almost poetic semihysteria that unofficially and informally allows deviant thinking and behaviour.'²⁴ Yoshio Taniguchi inverts the received opinion that incineration units ought to be hidden away by creating a dramatic towering show of them with his Naka Incineration Plant in Hiroshima City (pages 25 and 46–47). With Big Window House (pages 186–87), Tezuka Architects remove a great chunk of façade so that discarded shoes or a pair of dangling feet become the only markers delineating interior from exterior. In such a highly controlled society as Japan, the 'back-to-nature' theme, often used by Japanese architects to describe their own work, takes on a rebellious tone. Even Ito, with his tight grasp of cutting-edge technology, maintains that the ultimate aim for his curvaceous architecture is to reconnect us with nature (page 16).²⁵

Japanese deviationism reaches far into the past. Previously, the construction of any building in Japan involved a close one-to-one relationship between the master builder and his client, whose eccentric tastes attested to the 'deviations' found in many Japanese historical buildings all over the country, such as Kinkakuji (the Golden Pavilion), Katsura Detached Palace and Toshogu Shrine. Taste is, as Hajime Yatsuka points out, 'more than just a preference', rather 'a philosophy or vision of the world, a hermeneutic entity or a world of words'.²⁶ It seems not too far-fetched, then, to assume that when Japan catapulted itself into modernity, the client (or dreamer) fused with the master builder (the technician; even today, a qualifying architect must receive basic engineering training) to become the architect. So when Isozaki further comments that 'the elaboration of details' was left to the judgement, whim and intuition of the master builder, we can argue that architects in Japan simply refused to fit the norm; or rather, that deviation has, in this new discipline, become the norm.

According to the architectural historian and architect Terunobu Fujimori, tea masters generally acted as both client and engineer of their own tea houses, preferring not to involve a skilled carpenter for fear of being too ostentatious.²⁷ The space was kept deliberately small and simple, shutting out the outside world with *shoji* screens so that the master could concentrate on the performance of making tea in front of the guests. Comments about *kakejiku* (picture scrolls), *ikebana* (flowers), *utsuwa* (tableware) and, finally, the taste of the tea would follow in a highly ritualized manner. The hut itself merely served as a

The users of Kaito Workshop by Junya Ishigami + Associates are clearly visible from outside.

With the help of the structural engineer Mutsuro Sasaki, Toyo Ito creates a billowing roof for Meiso no Mori Municipal Funeral Hall.

Rooftop view of Hanamidori Cultural Centre, Tokyo, by Atelier Bow-Wow – a park on top of a building.

container, but it was in this space that 'the building, objects, fire, bodies and words' mingled to produce an intense experience.[28] Fujimori's own tea house, Takasugi-an (pages 10 and 122–23), a twist on this traditional building type, seems to dissipate this somewhat stifling intensity through its wide-open window. Fujimori's approach is typically postmodern as he takes history with a pinch of salt.

As the founder of Kenchiku Tanteidan (Architecture Detectives) in the 1970s and subsequently an active member of ROJO (Roadway Observation Society), which began in the mid-1980s, Fujimori was one of the first to embrace the 'deviations' found in Japanese cities. According to Fujimori, ROJO is committed to searching out, photographing and labelling 'objects' found in rambles around the city, such as *jun-kaidan*, 'the pure staircase' that has somehow survived demolition and now leads nowhere.[29] By bringing people's attention to such curiosities, the group has made Japan's cities more endearing and accessible. Its work, moreover, has made a strong impression on the younger generation of architects, among them Yoshiharu Tsukamoto and Momoyo Kaijima of Atelier Bow-Wow. Tsukamoto and Kaijima surveyed a series of *dame* (no-good) architecture – such as the Super Car School, a mammoth concrete block containing a supermarket and driving school, with a mock city for learners to drive around on its roof – in their book *Made in Tokyo* (2001).

Atelier Bow-Wow broke the mould by fully exploiting Tokyo's cramped and chaotic urban context. By giving its Gae House (pages 192–93) an enormous roof and protruding eaves, it subverted the strict building-to-site ratio imposed by the authorities, by which the available land surface was reduced to a mere 80 square metres (860 sq. ft); it also increased the volume of, as well as bringing light into, the top floor with horizontal glass panels tucked under the eaves. In the suburb of Tachikawa, its Hanamidori Cultural Centre (left; pages 38–39), in an echo of its unlikely 'prototype', the Super Car School,

16 NEW ARCHITECTURE IN JAPAN

carries a re-creation of the surrounding parkland on its rooftop, complete with street lamps, while lifts and escalators unsettle the visual simulacrum. The crucial difference here of course is that the driving school's hybridity was an afterthought rather than an integral part of the design. The embracing of context, then (whether urban or non-urban), disguised at times simply as a protest against policing and regulations, plays a conspicuous role in Atelier Bow-Wow's serious work.

Occasionally a fresh pair of eyes (such as those of a foreign architect) is needed to see what context there is. Tokyo's complex zoning laws, which fill the city's air with an invisible matrix of lines and restrictions, led the architects Herzog & de Meuron to carve their notional Prada store into its final angular form (pages 256–57). The architects' radicalism comes, however, not so much from the iconic building they constructed as from the structure they did not build: they left part of the site empty, creating a small pocket of air in the hectic city. According to the architects, it is supposed to function in the same way as European plazas, 'creating a sense of luxury'. Under Tokyo's high-rise skyline, criss-crossed with electric wires, however, it acts as much as a negative space, in which secrets and hidden desires are to be invested, like the space under a table or chair. Interestingly, Isozaki points out the lack of Western-style open plazas in old Japan. What we find instead is *kaiwai*, a more temporary, amorphous area where festivals and other communal events take place. Crucially, then, activities loosely define *kaiwai*.

The 'street' in Sou Fujimoto's Children's Centre for Psychiatric Rehabilitation in Hokkaido (above; pages 176–77) is another kind of negative space. The centre sits in rolling countryside, but its appearance is that of a congested neighbourhood extracted directly from Tokyo and relocated without disturbing its jumble of structures. The street is made by loosening a cluster of buildings and reconnecting the gaps to form a continuous communal space that meanders through the cluster, sometimes narrowing, sometimes expanding, and finally forking into alcoves and corners that serve no particular function, apart from perhaps piquing the children's imagination. Fujimoto's street, then, is defined by the physical as well as the mental 'activities' of the children.

In the mid-1920s the architect Sutemi Horiguchi called his approach to design *hi-toshiteki-na-mono* (non-urban elements).[30] 'Non-urban elements' seem just as important for Fujimoto as he taps into the 'primitive' energy buried inside us. The classical cuboid structure of his Final Wooden House in southern Kyushu (pages 100–101) disguises an interior of chunky timber planks, rough-cut and densely packed. Users must negotiate oversized steps and hanging pieces of timber, as though in a forest. The bungalow thus preserves the 'wildness' of the surrounding forest, and – indirectly – the quality Fujimoto finds in the distant, pulsating city of Tokyo.

Isozaki identifies 'two kinds of Japan-ness': the earthy, more dynamic or 'populist' aesthetic of the Jomon period (exemplified by pots with pressed rope patterns), which is said to date to as far back as 8000 BC, and the more elegant, even 'aristocratic' or 'elitist' aesthetic of the Yayoi period (exemplified by

Sou Fujimoto Architects' Children's Centre for Psychiatric Rehabilitation in Hokkaido consists of a cluster of cuboids, the layout of which seems to have been modelled on an imaginary city centre.

The strange and wonderful Community Hall inside Aomori Museum of Art by Jun Aoki & Associates.

haniwa figurines), which dates back to AD 300.³¹ It is a dialectic that has been operating in Japan for some time; according to Isozaki, some architects – in particular, Kenzo Tange with his Peace Centre in Hiroshima City – have suffered in the post-war democratic, anti-colonial climate from having their work identified with 'elitist' Yayoi-esque aesthetics. While one must avoid pigeonholing any architect in this way, it is remarkable that this Jomon-versus-Yayoi duality, or these 'two kinds of Japan-ness', seem still to apply: the jagged, earthy work of Fujimoto and Fujimori on one side; and the smooth, transparent, more refined output of Ito, Kengo Kuma and SANAA on the other.

Jun Aoki's Aomori Museum of Art, on the northern tip of Honshu (pages 62–63), seems to display both personalities. The competition-winning scheme caps a system of 'trenches', some of which are 20 metres (66 ft) deep, with a snow-white mass of steel and concrete containing workshops, cafes and offices. The trenches, which contain the main exhibition halls, expose the earth and directly refer to the Sannai Maruyama site, one of the largest and oldest Jomon-period archaeological sites in Japan, which is next to the museum. The rough texture created by the earth in the museum's lower level contrasts sharply with the smooth, white upper level. The heaviness and darkness of Aoki's subterranean world are also carefully offset by the whimsical addition of ornamental details, such as the white-curtained, arched windows (which would not be out of place on the façade of an Edwardian house) and the playful Community Hall.³²

Jomon or Yayoi: whatever the lineage, the beauty of this book ultimately comes from the unity of its images, all taken by one photographer. His gaze is, of course, external (a fresh pair of eyes), for he is British, but – crucially – it is also impartial. The photographs are not the 'official' versions, taken hastily soon after the buildings are completed, empty and devoid of context. Those images, often heavily worked on, are usually pallid and lifeless. Such representations seem wholly inappropriate here, for (as we have seen) architecture is the stuff of dreams as well as of function. Dreams are the visions that these architects hold close to their hearts. Moreover, Japanese spatial renderings, whether contemporary or historical, are inherently performative; that is to say, they are incomplete without people. For this reason, they deserve close scrutiny. The American writer Susan Sontag wrote in her seminal book *On Photography*: 'To take a photograph is to participate in another person's (or thing's) mortality, vulnerability and mutability.'³³ Only when we see these buildings in use can we begin to discover what is really Japanese about them.

1. Roland Barthes, trans. Richard Howard, *The Eiffel Tower and Other Mythologies* (Berkeley, Calif.: University of California Press) 1997, p. 6.
2. Kenneth Frampton, *Modern Architecture: A Critical History* (London: Thames & Hudson) 2007, p. 281.
3. Ibid.
4. The Tower, due for demolition, will sadly belong once again to the immateriality of dreams.
5. 'For us Isozaki without words is like Le Corbusier without words.' Hajime Yatsuka, 'Autobiography of a Patricide: Arata Isozaki's Initiation into Postmodernism', *AA Files* 58 (2009), p. 68.
6. Arata Isozaki, trans. Sabu Kohso, *Japan-ness in Architecture* (Cambridge, Mass.: MIT Press) 2006, p. 65.
7. Marius B. Jansen, *The Making of Modern Japan* (Cambridge, Mass., and London: Harvard University Press) 2000, p. 588.
8. Notoriously, one lane forks into a tunnel between the fifth and seventh floors of a tall commercial building, for which the road company, as a 'tenant', must pay rent.
9. Ando's slight openings remind me of the slits made for the downcast eyes on the statues of Kannon Bosatsu, an enlightened being who, unlike the Buddha, decided to remain on Earth for the well-being of others. Sometimes these statues are so enormous that one can go inside them, enabling a unique view of the world.
10. I am indebted to the London-based Japanese architect Takero Shimazaki, who gave a highly personal and insightful talk about the Japanese influence on his own work in the forum 'Transcending Boundaries: Concerning Contemporary Architecture in Japan' at the Japanese Embassy in London on 22 April 2009. In his talk, he coined the term 'loose-fit', which succinctly summarized the recent trend for me.
11. In the past, Japanese houses have been compared to 'rabbit hutches' because of their compactness; Kurokawa, who died in 2007, was therefore being sarcastic when he referred to himself as a wild rabbit (the quotation is from a conversation I had with Kisho Kurokawa in the spring of 2005).
12. Yuki Sumner, 'Interview: Sou Fujimoto', *Blueprint* 264 (12 February 2008), p. 44.
13. Yukio Mishima was a renowned Japanese writer who committed ritual suicide by *seppuku* (disembowelling) inside an army headquarters in 1970, calling for a stronger and purer Japan through the restoration of political and military power to the Japanese emperor.
14. Perhaps I write too hastily that the Hiroshima Genbaku Dome is the only ruin in Japan, but most historical buildings (castles and temples) in the country are faithful copies of the originals. Such structures are very rarely left as ruins.
15. Kenzaburo Oe, Nobel Lecture, 7 December 1994, nobelprize.org/nobel_prizes/literature/laureates/1994/oe-lecture.html.
16. Isozaki, *op. cit.*, p. 65.
17. Oe, *op. cit.*
18. Kristin Feireiss, 'An Interview with Kazuyo Sejima and Ryue Nishizawa', in *Kazuyo Sejima + Ryue Nishizawa, SANAA: The Zollverein School of Management and Design, Essen, Germany*, ed. Kristin Feireiss (Munich: Prestel) 2006, p. 64.
19. Thomas Daniell, *After the Crash: Architecture in Post-Bubble Japan* (New York: Princeton Architectural Press) 2008, p. 105.
20. It is hard to believe now that in the late 1990s it took nearly six years of debate to convince various groups of people, including librarians and curators, of the wisdom of his radical idea.
21. To give SANAA credit, it has tried to resolve the conflict between the practical aspects of running the museum and its own vision by introducing flexible, wall-like sliding doors that screen private areas discreetly. The partitions are painted white (in fact, everything in the museum is white, even the furniture).
22. The museum is dedicated to the work of Tomihiro Hoshino, an artist who was paralysed from the neck down in an accident and who now uses his mouth to paint.
23. 'In Japan one's identity seems to be defined to a greater extent by the social environment and one is responsible to that environment only. This could be a motorcycle company, a baseball team, a theatre group, or even the entire country; all depending on the time and place. The point is that one cannot really exist separately from these groups without risking severe psychological problems …. Alienation in Japan is a very high price to pay. Not conforming to the expected pattern means essentially that one does not exist at all.' Ian Buruma, *A Japanese Mirror: Heroes and Villains of Japanese Culture* (London: Jonathan Cape) 1984, p. 100.
24. Alan Williamson, 'Koan Ranger', interview with Gary Snyder, Poetry Foundation, 29 May 2009 (poetryfoundation.org/journal/article.html?id=181667).
25. Conversation with Toyo Ito, Spring 2008.
26. Yatsuka, *op. cit.*, p. 70.
27. Tea making was ritualized as a genteel activity in Japan roughly four hundred years ago.
28. Terunobu Fujimori, *Fujimori Terunobu Kenchiku* (Tokyo: TOTO Shuppan) 2007, p. 50 (my own translation).
29. Jordan Sand, 'Street Observation Science and the Tokyo Economic Bubble, 1986–1999' in Gyan Prakash and Kevin Michael Kruse (eds), *The Spaces of the Modern City: Imaginaries, Politics, and Everyday Life* (Princeton, NJ: Princeton University Press) 2008, p. 382.
30. Isozaki, *op. cit.*, p. 11.
31. Ibid., pp. 33–46.
32. The Community Hall is a room that seems to serve no purpose other than to create intrigue. Situated at the core of the museum, its walls are exact replicas of one another (even down to the doors), so that visitors suddenly find themselves at a loss as to how to emerge. Aoki plays the artist here, demanding of his visitors certain artistic sufferings ('Where am I? How do I get out?'), rather than simply being a supplier of functional space.
33. Susan Sontag, *On Photography* (London: Penguin Classics) 2002, p. 15.

Architecture in Japan: In Context
Naomi Pollock

In September 1988 I arrived at Tokyo's Narita International Airport, my Mayline straight edge stowed in my luggage and my curiosity at the ready. A practising architect from New York, I came as a guest of the Japanese government, which awarded me a generous scholarship and the opportunity to become a graduate student once again. I turned out to be a good investment: as an architect-turned-journalist, I have been spreading the word about Japanese design for the better part of the past twenty years.

Living in Tokyo for the bulk of that time has enabled me to watch the changing architectural scene from within. Through this immersion I have acquired an in-depth understanding of Japan's unique built environment and the forces that mould it. I have had the time to visit and then revisit building sites. I have had the opportunity to investigate, right down to the details. And I have had the chance to talk to architects, engineers and contractors, as well as clients.

This collective body of knowledge makes it possible to explain Japanese architecture in context – itself rather an enigma. Unlike in the West, context in Japan is a somewhat abstract concept that needs to be defined broadly, not just in terms of simple adjacencies. At that level Japanese cities do not make much sense, since their buildings are created with little regard for stylistic compatibility or harmonious relationships. This is especially true of central Tokyo, where land is extremely limited.

The conspicuous absence of such urban organizational devices as discernible skylines, well-defined waterfronts, consistent street grids and green belts compounds the visual chaos. While communities and neighbourhoods have developed organically around major intersections and transportation hubs, Japanese cities tend to be arranged concentrically. Municipalities are divided into wards made up of districts comprised of multi-sided blocks of buildings numbered in the order of completion. As a result, building 7 stands next to building 42, which stands next to building 3, and so on. Against this backdrop, it is no wonder that buildings are treated and seen as individual objects even if their neighbours are only centimetres away.

One of the benefits, as well as deficits, of this system is that individual buildings are replaced at the drop of a hat. In Japan, private homes stand for no more than thirty years. Although large commercial and institutional projects may survive for longer, trendy restaurants and shops seldom remain more than five years, in keeping with the country's seemingly insatiable appetite for the new. As a result, there is little impetus to engage in an aesthetic or formal dialogue with the neighbours. More often than not, this ephemerality breeds an attitude of detachment and disengagement.

For some architects, the solution is simply to ignore the proximate surroundings; others adopt a more aggressive approach by putting up unfriendly, view-blocking walls. While focusing attention inwards, this strategy also edits out unsightly neighbours and protects against any that might appear in the future. Although effective, this defensive approach all but severs the tie between the goings on inside and the activity outside.

Yet among younger designers there is a growing acceptance of, and even a willingness to embrace, the Japanese urban condition. When the principals of Atelier Bow-Wow decided to build their combined home and office in the heart of Tokyo, they happily bought a flag-shaped site hemmed in tightly by buildings on all sides. Totally at peace with this rather extreme, but not uncommon, condition, the architects welcomed the man-made scenery with huge windows and spacious porches, visually incorporating the stucco walls, concrete-lined crevices and discordant roofs near by. Instead of turning their backs on the cityscape, the architects allow it to waft through their home like a gentle breeze (overleaf; pages 248–49).

Tod's Omotesando Building in Tokyo by Toyo Ito & Associates, Architects, exemplifies the value placed on good commercial design.

But this receptive attitude towards the city and the design problems it presents extends beyond residential work. In part it is also a function of the available jobs. In recent years commissions for flagship stores and high-end boutiques embedded in the urban fabric have surpassed those for object-like museums and cultural facilities, which were sought after in the 1970s and 1980s. These commercial developments have given facelifts to major thoroughfares and elegant boulevards, but they also indicate the economic value assigned to good design: buildings produced by creative architects attract attention and enhance brand image in the eyes of the Japanese consumer. What Toyo Ito's Tod's headquarters does for shoes and bags in the middle of Tokyo (pages 22 and 262–63), Akihisa Hirata's Showroom H does for lawnmowers and snow blowers in the wilds of Niigata Prefecture (pages 260–61).

This perception of design as a revenue-generating commodity began in earnest in Japan during the country's economic bubble of the late 1980s. At that time Japan began importing world-famous architects in droves in the hope that the cachet of an Aldo Rossi hotel would lure visitors to Fukuoka, or that Philippe Starck's Tokyo headquarters for Asahi Breweries might boost beer sales (or at least update the image of the century-old company). Situated alongside the Sumida River, the latter's bold black volume capped by a flame-shaped gold protuberance mystified architectural audiences at the time of its completion in 1989, and for months the eye-popping building was the talk of the town.

Many of the so-called 'bubble buildings' – not just those designed by foreign firms from afar – were insensitively designed or poorly executed. Yet their presence raised the profile of architects and architecture across the country, and this perception did not change even once the money had dried up.

On the contrary, government agencies of all levels began commissioning architects to design public works projects in the hope of reinvigorating the failing economy. Some of the resulting buildings were ill-conceived, and ended up as under-utilized *hakomono* (empty boxes). This was especially true of museums: many were beautifully designed but with little regard for long-range curatorial vision, budgets or storage for growing collections. Another problem was the routine reshuffling of government officials, which meant that architects had to cope with an ever-changing clientele as each new appointee put his or her imprimatur on the project. But many of the completed projects substantially ratcheted up the quality of existing facilities or improved the infrastructure by providing

Atelier Bow-Wow's House and Office in Tokyo makes clever use of an awkward site.

Naka Incineration Plant (left) by Yoshio Taniguchi brings this usually hidden function into the public domain.

Jun Igarashi Architects' Ware House (right) is a private viewing gallery for classic cars.

libraries, community centres, concert halls and other amenities that had been missing.

In an effort to promote transparency in the selection process and to encourage participation by a wider range of designers, many government agencies selected architects through competitions. This was the case with the Yokohama International Port Terminal, completed by the London firm Foreign Office Architects in 1994. Elsewhere, architects were invited directly or through such large-scale initiatives as Kumamoto Prefecture's Artpolis programme, modelled on the Internationale Bauausstellung in Berlin.

Artpolis (begun in 1988) was the brainchild of the then governor of Kumamoto Prefecture (and later prime minister of Japan), Morihiro Hosokawa, and his advisor Arata Isozaki, the established Tokyo-based architect, also from Kyushu. Under Isozaki's aegis, the programme identified a group of inventive architects and invited them to create blocks of flats, museums, bridges and other public works projects around the prefecture. Artpolis's mission was to propose an alternative method of awarding government-sponsored work, and to break the cycle whereby the same architectural firms and construction-company design departments were hired over and over again. At the same time, it successfully transformed a remote agricultural prefecture into an architectural hotspot.

Unsurprisingly, there were some hiccups as Artpolis's first phase drew near to completion. Although architecture critics praised the outcome, some local designers felt snubbed by the programme's parachuting in architects from outside, and some Kumamoto citizens questioned the suitability of the often avant-garde buildings. But one indication of Artpolis's success was the similar initiatives it spawned in other parts of the country, including Hiroshima 2045: City of Peace and Creativity. Launched in 1995 to mark the fiftieth anniversary of the atomic bomb, the programme awarded government-sponsored projects to such famous architects as Yoshio Taniguchi, whose Naka Incineration Plant (above) is one of the world's most elegant waste-disposal facilities.

But the positive impact of such programmes as Artpolis was not just that they elevated the design value of individual buildings or cast a spotlight on regional locales. As in the Bubble Period's earlier building frenzy, Artpolis created opportunities for many young architects whose lack of experience would have previously disqualified them. Before the economic boom, budding architects tended to enter graduate courses at one of the country's prestigious universities, and started their own practices only after completing lengthy apprenticeships in established firms. This was deemed necessary for training and was also a valuable source of referrals, since mentors passed jobs to their protégés. But during the Bubble Period, newly minted designers – and even graduate students – were given and leapt at chances to build immediately. And once the system changed it never reverted: instead, it paved the way for such designers as Sou Fujimoto, who, eschewing apprenticeship altogether, pursued his own conceptual work straight out of university until he was given the chance to build mental-health facilities for psychiatrists (among them, his father) in Hokkaido.

Façade detail from Leaf Chapel by Klein Dytham Architecture.

21_21 Design Sight, Tokyo, by Tadao Ando Architect & Associates.

In part because of the deeply entrenched education and training system before the Bubble Period, design-orientated architects are still concentrated in major cities, such as Tokyo and Osaka. Although many of the works completed around the country during the 1980s and 1990s were the products of those firms, the geographic dissemination of architecture readied the regions for the emergence of young designers on their home turf. And in recent years innovative firms have begun to crop up in Fukuoka, Hiroshima, Kyoto, Sendai and other mid-sized cities on the Japanese archipelago.

Part of the appeal of these designers is their familiarity with the local climate, lifestyle and building culture, since Japan, despite being a small nation, spans a wide range of weather extremes and construction practices. When a collector of classic cars decided to build a private gallery in the middle of Hokkaido, architect Jun Igarashi was the clear frontrunner for the job (pages 25 and 124–27). A native of the northern island, the designer came equipped with first-hand knowledge of the area's temperature ranges and snowfall. And instead of reinventing the wheel, he knew to take his design cues from the masonry storehouses that are a centuries-old Hokkaido tradition.

But thanks to the proliferation of design magazines and the ever-expanding Internet, even such isolated architects as Igarashi obtain commissions all over Japan. The one place where few architects have major jobs (aside from residential work) is Tokyo: because of the city's crowded condition, large open lots are hard to come by. Yet that has not stopped developers from planning giant projects, such as Roppongi Hills and Tokyo Midtown, with looming towers that contrast dramatically with the small-scale buildings and tiny streets near by.

While Roppongi Hills was built on land amassed from small parcels purchased over many years, Tokyo Midtown was an exceptionally rare case that came about when a large property belonging to the Defence

Natural Strips II by Masaki Endoh/EDH adds another layer to Tokyo's complex urban fabric.

Agency was sold off all at once. Located within walking distance of each other, the two mixed-use complexes both include offices, apartments, luxury hotels and high-end shops. They also contain public gardens and museums that have improved the quality of their neighbourhoods: Kengo Kuma's Suntory Museum of Art (pages 86–87) is embedded in Midtown's shopping centre, Tadao Ando's 21_21 Design Sight (opposite, bottom; pages 58–59) sits within its park, and Gluckman Mayner Architects' Mori Art Museum caps the fifty-four-storey centrepiece of Roppongi Hills.

Designed respectively by the New York firms Kohn Pedersen Fox Associates and Skidmore, Owings & Merrill, neither the skyscrapers of Roppongi Hills nor the high-rises of Tokyo Midtown are distinctively Japanese. In fact, both complexes would be at home in many big cities around the globe. Perhaps that is their intention. Despite these and other attempts at internationalization, Japan remains a remarkably insular place, where few foreign architects have been able to get a strong foothold. One notable exception is Klein Dytham Architecture, whose imaginative projects – such as the trio of interventions at a resort in Yamanashi Prefecture (opposite, top; pages 96–99, 112–13 and 178–79) – have charmed clients and critics alike. But beneath the gloss of mirror finish and whimsical wallpaper, the practice's works are the sort of friendly commentaries on contemporary Japan that only outsiders have the detachment to make.

Nowadays Japan borrows selectively from the West, usually hiring overseas architects for a particular expertise, such as skyscraper design or the planning of such healthcare facilities as nursing homes, which are new to Japan but have existed overseas for some time. Otherwise, fresh crops of home-grown designers have been turning the country into an incubator for innovation and invention that is envied and closely watched by the worldwide architectural community. 'Until about twenty years ago, you often heard Japanese culture being dismissed as "all copies, no originals" …. It's now an indisputable fact that Japan has become a nation that exports culture', writes Hitoshi Abe, the Sendai architect imported by the University of California in 2007 to chair its Department of Architecture and Urban Design.[1]

Freed from the historical and aesthetic constraints facing their counterparts in the West, architects in Japan readily experiment with new structural systems, construction materials and geometric solutions. Although subject to tough legal restrictions (especially in Tokyo) and deeply entrenched social conventions, most design-orientated firms treat those limitations as catalysts for, not impediments to, good design. And, as this book attests, the fruits of their labours are rich and diverse. Within this 'anything goes' atmosphere, nothing really stands out. A glass box covered with metal fins, Shuhei Endo's Natural Strips II (above; pages 214–15) may seem bold and brazen to Western eyes. But in Tokyo it is just another variation of normal.

1 Hitoshi Abe, 'Study of the Edge', in Thomas Daniell, *After the Crash: Architecture in Post-Bubble Japan* (New York: Princeton Architectural Press) 2008, p. 8.

30 Centrair
32 Chokkura Plaza
34 Grin Grin
36 Halftecture OJ
   Halftecture OO
   Halftecture OR
38 Hanamidori Cultural Centre
40 Hoshakuji Station
42 Meiso no Mori Municipal Funeral Hall
46 Naka Incineration Plant
48 Naoshima Ferry Terminal
50 Slowtecture M
52 Tokyu Toyoko-Line Shibuya Station
54 Yokohama International Port Terminal

Infrastructure and Public Spaces

Centrair

Ise Bay, Tokoname, Aichi Prefecture

NIKKEN SEKKEI AND HOK, 1998–2005

BELOW, TOP The airport is built on an artificial island. BELOW, CENTRE AND BOTTOM Built away from residential areas, Centrair operates twenty-four hours a day. OPPOSITE Large trusses create column-free spaces so that even mature trees can grow inside the building.

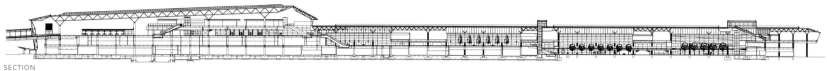

SECTION

Built on more than 405 hectares (1000 acres) of artificial island, this new international airport – formally known as Central Japan International Airport – serves central Japan and can handle 20 million passengers a year (12 million domestic and 8 million international). The transportation of manufactured goods is also a key function: the region, which is home to the car-maker Toyota, is heavily industrialized, and around 600,000 tonnes of cargo pass through the airport annually.

Designed jointly by Japanese firm Nikken Sekkei and US practice HOK, the main building is about 1030 metres (3380 ft) long and 500 metres (1640 ft) wide. Configured as a T-shape, intended to echo that of a crane (a symbolic bird in Japan) spreading its wings, the building comprises a central terminal from which three large concourses project. A roof of 80,000 square metres (860,000 sq. ft) covers the building, which itself contains 220,000 square metres (2,370,000 sq. ft) of floor space. Since it opened, however, the airport has already been expanded, mainly with new cargo-handling facilities; plans are also afoot to construct a second runway 300 metres (980 ft) from the existing one.

Centrair is designed for super-efficiency. Set in a marine location well away from any dense residential areas, it can operate twenty-four hours a day; transport links are provided by a four-lane motorway, a railway line and even a high-speed boat service. Within the building itself, services are provided in the central hub, while international and domestic flights are separated into distinct zones. To reduce the confusion inherent in many large airports, facilities for arriving and departing passengers are on different levels. Large trusses and the minimal use of columns combine to create wide, open spaces, and plentiful glazing (modulated by metal louvres and screens) admits enough natural light to prevent the interior from becoming gloomy. There is even sufficient daylight to allow a line of mature trees to thrive inside the building.

Apart from the sheer length of the glass façade visible on the approach to the building, arguably the most distinctive architectural feature is the roof treatment. Not only is the building's steel structure on show, but also a virtue has been made of its slanting forms to create a composition of folds and folds within folds. Its faceted treatment gives the structure upward mobility, as though it is ready for take-off at any time. Centrair is now Japan's third-largest international airport, after Narita and Kansai.

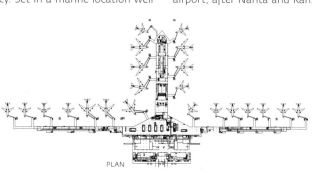

PLAN

INFRASTRUCTURE AND PUBLIC SPACES 31

Chokkura Plaza

Takanezawa, Tochigi Prefecture

KENGO KUMA & ASSOCIATES, 2006

This single-storey multi-purpose building and exhibition facility is an exploration of the possibilities inherent in a particular type of stone. Built of a very soft, porous rock called *oya*, it provides an alternative vision for the contemporary railway architecture typically characterized by steel and glass.

Kuma first became interested in an old, disused rice warehouse across the plaza, constructed from oya. Used by Frank Lloyd Wright in the design of the Imperial Hotel in Tokyo, oya contains holes and indentations in which dust and soil collect, forming a mottled appearance of brown, khaki and grey. The deep-brown portion of the stone is often referred to as *miso*, for its similarity to the paste used to make miso soup. For Kuma, this textured stone offered the chance to design a building that was clearly *of* the local landscape, rather than just occupying space *in* it. 'I thought it should be a dignified building that takes deep root in this locale,' he explains.

In Chokkura Plaza, the stone is used in such a way that it appears to form both cladding and structure. Although the frame and soffits of the building are steel, the stone is applied not as a thin surface layer but in forms of real depth. Its 'sponge-like' quality is explored, leading to the distinctive weave of the façade, composed of a grid of solid and void. The entire building, which rises to a height of 8.2 metres (27 ft), resembles a piece of basketry or even a textile.

The 670-square-metre (7210-sq.-ft) building, next to Hoshakuji Station (also refurbished by Kuma; see pages 40–41), is extremely simple in plan, divided almost equally into three principal spaces: cafe, souvenir shop and small exhibition hall. One section of the building opens and acts as a shelter, with steel latticework on the ceiling echoing the patterns on the façade and gently encasing the void. The rice warehouse itself has been renovated and reinvented as a conference and assembly centre.

In his book *Japan-ness in Architecture* (2006), Arata Isozaki points out the lack of Western-style plazas in old Japan. Instead, he writes, what we find is a more ambiguous demarcation called *kaiwai*, a temporary and amorphous area where festivals or other communal events take place. Despite the more modern suggestion of its name, Kuma's plaza retains some of the traditional sensibility of *kaiwai* by being fluid and, most of all, open and inviting.

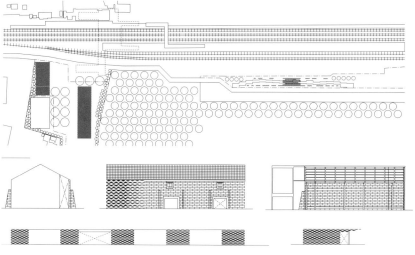

SITE PLAN AND ELEVATIONS

OPPOSITE Chokkura Plaza stands next to Hoshakuji Station, also designed by Kuma. TOP LEFT AND BOTTOM RIGHT The use of oya cladding produces a distinctive appearance. TOP RIGHT Steel latticework on the ceiling echoes the patterns on the façade. BOTTOM LEFT Sandwiched between glazed curtain walls, the cafe also participates in the interplay of solid and void.

Grin Grin

Fukuoka City,
Fukuoka Prefecture

TOYO ITO & ASSOCIATES, ARCHITECTS,
2002–2005

This undulating set of landform spaces, configured like a set of overlapping petals, is in Island City Central Park, on a man-made island in Hakata Bay in the port city of Fukuoka. Space was reserved on the 400-hectare (988½-acre) site – with container ports, a business district and residential areas all sparsely laid out – for a park centred on a lake. The curving forms and sinuous lines of the park and of Grin Grin, within it, are the perfect counterpoint to the more formal development elsewhere on the island.

The lakeside building is an educational, as well as recreational, centre, which celebrates the flora of the region. Grin Grin also contains an office, but a large part of it is dedicated to the display of tropical plants, so it effectively functions as a greenhouse with a grass-covered roof. Yet another collaboration of Toyo Ito with engineer Mutsuro Sasaki, it is composed of a single continuous surface that twists at two points to form three interior spaces.

The ribbon-like roof-form, 400 millimetres (15¾ in.) thick and 185 metres (607 ft) long, achieves column-free spans of 50 metres (164 ft) and is pierced by four apertures, flooding the interiors with natural light. Paths wind around, through and over the structure, providing a sinuous journey and offering glimpses through the glazed roof lights into the plant-filled spaces below.

The route is deliberately circuitous, in the manner of the picturesque walks in eighteenth-century English gardens. Echoing a younger generation of architects, such as Sou Fujimoto, Sasaki bemoans the extreme homogenization and optimization of contemporary architecture, which creates 'an excess of convenience that has weakened modern people physically and mentally'.

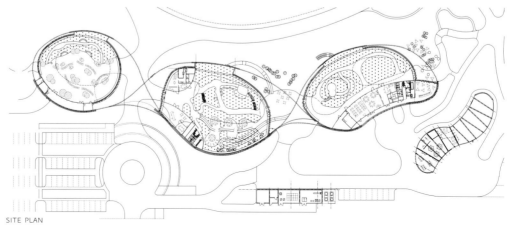

SITE PLAN

Through such a daring scheme as this, he would like us instead to 'reclaim a basic biological instinct: namely, taking care when walking'.

With an area of more than 5000 square metres (53,820 sq. ft), this building has a claim to be more than just a greenhouse. The green rooftop does not mimic parkland, as the Hanamidori Cultural Centre (which Ito designed with Atelier Bow-Wow; see pages 38–39) does in the Tokyo suburbs; and much of the building will eventually be covered in vegetation. It represents, therefore, a piece of wilderness, something that is otherwise distinctly lacking in this 'brave new world' development.

OPPOSITE, TOP The configuration is reminiscent of a set of overlapping petals. OPPOSITE, BOTTOM The landform building is gradually becoming covered in vegetation. THIS PAGE Visitors are encouraged to walk into, out of *and* over the building.

Halftecture OJ, 2004–2005/Halftecture OO, 2004–2005/Halftecture OR, 2004–2006
Osaka City, Osaka Prefecture

ENDO SHUHEI ARCHITECT INSTITUTE

According to the Osaka-based designer Shuhei Endo, gravity is not a force to be fought but one to be worked with. Although most architects go to great lengths to combat its pull, Endo relies on its strength to snap his steel buildings into place. Commissioned by the city, this trio of public toilets plus a cafe at the base of Osaka Castle is a case in point.

Belonging to Endo's Halftecture as well as his Gravitecture series, each of these minimally enclosed projects was designed to take advantage of the scenery: a majestic castle with a tree-lined moat. Although the building was bombed in the Second World War and then reconstructed, its magnificent foundation walls have stood for hundreds of years thanks to the force of compression holding their massive stones in place. With a polite nod to this feat of engineering, Endo's three buildings all use steel support systems, but each interacts with gravity in its own way.

The first to be completed was Halftecture OJ, a parallelogram-shaped public toilet divided down the middle into men's and women's areas by a free-standing partition with wall-mounted sinks on either side. While low-maintenance white-enamelled walls wrap each group of stalls, the building is draped in a sheet of rust-coated Cor-ten steel. Creased along its diagonal axis and again at either end, the metal cover droops under its own weight but is stabilized by this natural deformity.

Similar in size and function, Halftecture OO replaced an existing facility. This time Endo suspended the steel cover like a hammock from A-frame end supports made of matching Cor-ten steel. The roof warped under the influence of gravity, settling comfortably into a structurally sound catenary curve that hovers protectively above the sinks and stalls.

The third project, Halftecture OR, is a doughnut-shaped building with a forty-cover cafe and toilets separated from each other by a courtyard garden. Endo chose a circular plan, since it may be accessed from any direction. He ignored its dominant centre point, however, sliding the garden off to the side and enclosing the cafe in glass to direct attention outwards. Like the other projects on the site, it is topped with a Cor-ten-steel roof; this one is held up by clusters of skinny columns positioned to allow the roof to sag and rain to roll off, and angled to counter both horizontal and vertical forces.

Because of the sensitivities involved in working on a heritage site, Endo wanted his buildings to balance new and old. In covering each one with Cor-ten steel, he chose a modern material that only improves with age.

OPPOSITE, TOP Two views of Halftecture OJ with its folded steel cover. OPPOSITE, CENTRE Three views of Halftecture OO with its suspended steel cover. OPPOSITE, BOTTOM Three views of Halftecture OR. Its doughnut-shaped steel roof is supported by clusters of slender columns.

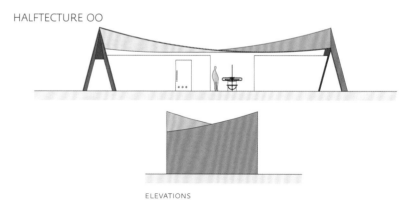

HALFTECTURE OJ

ELEVATIONS

HALFTECTURE OO

ELEVATIONS

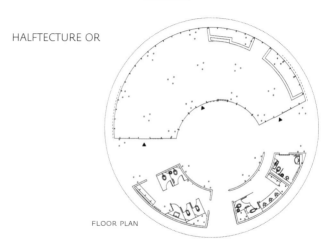

HALFTECTURE OR

FLOOR PLAN

INFRASTRUCTURE AND PUBLIC SPACES

Hanamidori Cultural Centre

Tachikawa, Tokyo

ATELIER BOW-WOW AND TOYO ITO & ASSOCIATES, ARCHITECTS, 2002–2005

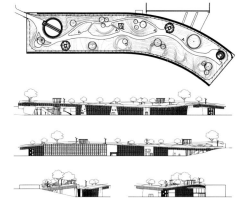

ROOF PLAN AND ELEVATIONS

SITE PLAN

In 2001 Momoyo Kaijima and Yoshiharu Tsukamoto of Atelier Bow-Wow published a book with Junzo Kuroda, *Made in Tokyo*, featuring many of Tokyo's most outrageous and absurd buildings: the 'centipede housing' under raised railway lines and the 'royal golf apartments', which offered a golf driving range on top of a tall block of flats, were among those they included. In line with this frantic hybridizing tradition in Tokyo, then, Hanamidori Cultural Centre, set in Tachikawa's spacious Showa Memorial Park, carries part of the park on its back. Momoyo Kaijima has coined the apt word 'parkitecture'.

The centre is one of a number of 'building-as-park' structures that are in vogue in Japan: Grin Grin in Fukuoka City, also by Toyo Ito (pages 34–35) and completed around the same time, is another. However, the Hanamidori Cultural Centre displays more of Atelier Bow-Wow's off-the-wall zest than Ito's ethereal touch, which is manifested here through elevator shafts resembling Sendai Mediatheque's famous columns, each one also an assembly of white steel tubes (pages 84–85).

An extraordinary illusion is created on the roof of the centre: a piece of parkland 150 metres long and 30 metres wide (492 by 98½ ft) – complete with mature trees, undulating hills, winding paths and even street lamps – appears simply to have been lifted off the ground. Only the people emerging from the escalators, which rise through a giant hole in the 'ground', or from the lifts, which pop up into mini-pavilions, betray the fact that one is on top of a man-made structure of steel and concrete.

Combining the words *hana* (flowers) and *midori* (greenery), the centre aims to educate about ecology and related sustainable topics. The centre is approached in two ways: over the roof and down via the escalators or lifts, or directly through the many entrances in the glazed façade tucked underneath. The long rectangular building, which assumes a broad curve at its northern end, uses more than just a green roof to integrate itself into the park: elements of the structure push through into the 'natural' realm above, tree-covered earthen mounds reflecting the position of conical meeting rooms and other facilities below.

The meandering interior, with nearly 5500 square metres (about 59,200 sq. ft) of floor space, embodies something of an airport or railway-station aesthetic. Carefully designed furniture partitions it into different areas, and strips of metal mesh hang like gauze from the ceiling to soften the lofty upper reaches, while trusses supporting the roof hint at branching tree trunks. Designed to accommodate exhibitions, workshops and other public events, the building can be opened to the park outside by folding away large portions of the façade.

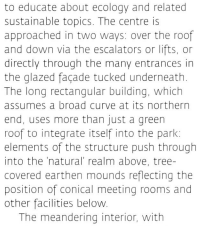

LEFT 'Parkitecture': a 150-metre-long (492 ft) tranche of parkland sits on the roof. OPPOSITE, TOP LEFT The glazed façade curves, like a hockey stick. OPPOSITE, TOP RIGHT Circulation cores push through the roof. OPPOSITE, BOTTOM The huge internal volumes are subdivided by flexible furniture installations.

INFRASTRUCTURE AND PUBLIC SPACES 39

Hoshakuji Station

Takanazawa, Tochigi Prefecture

KENGO KUMA & ASSOCIATES, 2005–2008

This railway station 130 kilometres (80 miles) north of Tokyo, developed for the East Japan Railway Company, demonstrates the art of making the ordinary extraordinary. Its form – a bridge flanked by covered stairways – is the norm for stations all over Japan. With a limited budget, Kengo Kuma was asked to embellish, rather than demolish and rebuild, the existing station, which stands next to Chokkura Plaza, his stone-and-steel exhibition space (pages 32–33).

Kuma approached the station project with discretion. The building is unexceptional from a distance, but the drama unfolds as commuters approach, and as they climb the stairs their eyes wander upwards to the timber stalactites that ornament this otherwise utilitarian structure. The plywood soffit hangs from the steel roof in undulating swathes, creating an unorthodox coffered ceiling of sharp angles. Slung in relatively shallow units in the principal upper-level spaces, these distinctive moulds drop down further nearer the edges of the structure, disguising the steelwork. Viewing the ceiling in motion, as one inevitably does, emphasizes the variation in form.

Like Chokkura Plaza, the new station exhibits a strong diamond motif: the former is defined by its walls of angular voids, while the diamond shapes at Hoshakuji Station, inspired by the diagrid engineering of the bridge, are more decorative. Approximately 1500 of these timber boxes are distributed across the ceilings of the station. Each box has the same base plan, but the angle and depth at which the projections are cut vary greatly. Although the shapes were plotted precisely with the aid of software, models and full-sized mock-ups were constructed to ensure the theory really did work in practice. The final pieces were fabricated off-site and each diamond form was fixed to the roof with steel hangers.

The result is a steel-and-concrete structure animated by a textured, and unexpected, inner surface. Kuma has said that transport architecture is typically uninspiring, and this ceiling treatment (which displays the architect's love of craft and tactility rather than of pure form) is his attempt to offer passengers a little more than just shelter.

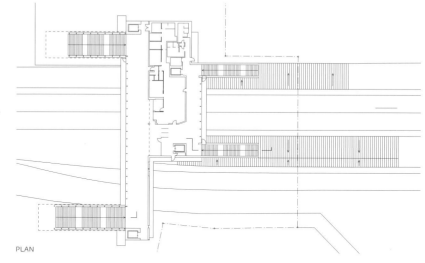

PLAN

OPPOSITE, TOP The station follows a typical bridge form found everywhere in Japan. OPPOSITE, BOTTOM The station seen from a platform. ABOVE The plywood soffit hangs in undulating swathes from the steel structure. RIGHT Functional and everyday space is lifted by the artistic ceiling. FAR RIGHT The boxes vary in depth and angle of cut.

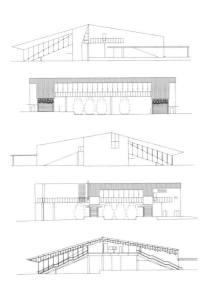

ELEVATIONS AND SECTIONS

INFRASTRUCTURE AND PUBLIC SPACES 41

Meiso no Mori Municipal Funeral Hall

Kakamigahara, Gifu Prefecture

TOYO ITO & ASSOCIATES, 2004–2006

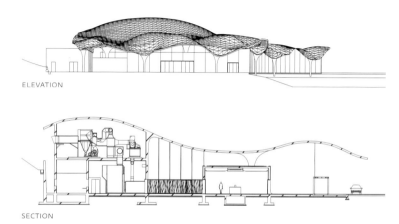

ELEVATION

SECTION

BELOW Main entrance. The concrete canopy echoes the hillside. OPPOSITE, TOP LEFT Rectilinear volumes lie beneath the billowing roof. OPPOSITE, TOP RIGHT The building wraps around one edge of a small lake. OPPOSITE, BOTTOM LEFT The canopy is softly lit from below. OPPOSITE, BOTTOM RIGHT A marble-clad interment room ensures privacy. OVERLEAF View at dusk, with the graveyard in the foreground.

In a curious way, this non-denominational funeral hall was constructed twice: once with the erection of the complex 'formwork' on which the concrete was poured; and again with the realization of the concrete itself. As with all concrete buildings, a mould has to be put in place first, but the bespoke timber canopy constructed for this project was so complex and well detailed that it could almost have stood as the building itself.

Echoing in its form the rolling hills of the surrounding countryside, the hall was envisaged by Toyo Ito as a place for quiet reflection, a space with sensuous, organic curves that would suggest a closeness to nature and the natural order of things. Nestled between wooded hills at the back and a large pond at the front, the hall has a 2270-square-metre (24,435-sq.-ft) roof that appears to hover in mid-air, as if it belongs neither to the world of the living nor to that of the dead.

The cutting-edge structural engineer Mutsuro Sasaki, who has enabled a number of projects for Toyo Ito, plotted the roof form on to a three-dimensional grid with 3700 node points. The formwork was created from timber and overlaid with a mat of steel reinforcing rods. Concrete was then sprayed over the mould to a depth of 200 millimetres (7⅞ in.) and the formwork removed, leaving the billowing concrete roof to support itself on beautifully tapered columns. With almost invisible floor-to-ceiling glazing, the anti-gravity effect is complete.

The structure curves around the pond. Inside, the undulating canopy is softly lit from below, and polished marble spreads like water over the floor. Unlike water, however, the same marble crawls up to clad the enclosed cubicles. A central corridor separates the valedictory and interment rooms from a furnace, which is completely hidden from view. Waiting and meeting rooms form a separate cluster of top-lit spaces under an extended wing of the canopy roof, benefiting from a clear view of the surrounding landscape; by contrast, the valedictory and interment rooms are completely enclosed for absolute privacy.

Ito's crematorium, with its almost free-form roof and the series of more logical spaces beneath, explores two architectural languages. Such duality poignantly makes way for the rituals of parting.

INFRASTRUCTURE AND PUBLIC SPACES 43

Naka Incineration Plant

Hiroshima City, Hiroshima Prefecture

TANIGUCHI AND ASSOCIATES, 1998–2004

As part of a city-wide regeneration programme called Hiroshima 2045: City of Peace and Creativity and implemented in 1995 'to create social infrastructure with superior design characteristics', such heavyweight architects as Riken Yamamoto and Hiroshi Hara were commissioned to work on various projects ranging from a fire station to municipal schools and parks.

The most innovative project undertaken under the auspices of the programme is the incineration plant designed by Yoshio Taniguchi. At the time it was commissioned, Taniguchi was highly regarded in Japan for designing a number of museums; he is now internationally acclaimed for his work on the extension of MoMA in New York, which was completed around the same time as this project.

Taniguchi is a master of configuration. When asked about his scheme for MoMA, he points out that his architecture aims not so much to create unique sculptural forms as to create the best environment 'where people can meet the artwork'. He shocked the museum's trustees by presenting the idea of telling the story backwards by starting the tours with the most contemporary works.

With Naka Incineration Plant, or 'Museum of Garbage' as he likes to call it, Taniguchi has confounded expectations in a similar way. The plant is poised on a reclaimed landfill site and runs on electricity generated by its highly efficient incineration system. A total floor space of 46,000 square metres (495,140 sq.ft) is accommodated in a bold structure clad in panels of lightweight galvanized steel and glass. A raised platform cuts through the seven-storey building and connects with the city's main avenue, which leads directly to the Peace Memorial Park to the north and the newly landscaped harbour to the south.

This platform, a 120-metre-long (394 ft) floating bridge, exposes the surprisingly pristine workings of the incinerator. With shining machinery towering on both sides, complemented by a magnificent view of the city at one end of the bridge and of the sea at the other, visitors cannot but be impressed by the link between the city, its waste and the natural environment into which it must be absorbed.

Discarding the received opinion that incineration units ought to be hidden away, Taniguchi invites us to admire, and also to acknowledge the inevitability of such a building in our cities. Its groundbreaking design means that this project incorporates both the beauty and the truth of modern urban living.

OPPOSITE Taniguchi exposes the incinerator's inner workings, which are normally hidden from public view. RIGHT The 120-metre-long (394 ft) bridge links the factory with the city and natural environment in a meaningful way. FAR RIGHT Gleaming machinery towers over the raised platform. BELOW The control room. BELOW, RIGHT Up to 4500 tonnes of household waste can be stored and processed in the refuse pit weekly.

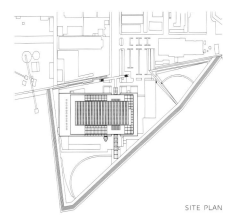

SITE PLAN

INFRASTRUCTURE AND PUBLIC SPACES 47

Naoshima Ferry Terminal

Naoshima, Kagawa Prefecture

KAZUYO SEJIMA + RYUE NISHIZAWA/SANAA, 2004–2006

Naoshima Ferry Terminal is the gateway to Naoshima, a small island between Honshu and Shikoku that has been revitalized by a splash of contemporary art. Naoshima Fukutake Art Museum Foundation has laid out a series of museums as well as various permanent art installations by notable artists, such as Tatsuo Miyajima and Hiroshi Sugimoto, in local villages. The terminal, the design of which was entrusted to the duo SANAA, signifies the island's transformation into an art site for incoming tourists.

The overhanging steel roof is the terminal's dominant feature, and provides shelter for people and vehicles. Raised far enough off the ground to accommodate high-sided trucks, the 70 x 52-metre (230 x 171-ft) roof is supported by a grid of extremely thin columns and eight 2.4-metre-long (8 ft) walls. These walls have mirrored surfaces, reinforcing the illusion that the roof plane floats unaided – or at least with the minimum of help. 'We have always been attracted by this ambivalence between something and nothing, by this floating of materials and space,' explain Kazuyo Sejima and Ryue Nishizawa in a booklet accompanying an exhibition of their work in the deSingel art gallery in Antwerp, Belgium, in 2007. 'It is an architecture that very generously gives people space and shelter, without thrusting itself upon them, without the intention of emphasizing the fact that it is architecture,' writes critic Moritz Küng in the same publication.

Nestled under the roof canopy is a glass box, inside which the cafe, tourist information office and waiting room are loosely cordoned off by low-level walls. Texture is reserved for the underside of the roof in the corrugations of the cladding, while the glazing seems to exist simply to trap light, and be nothing else.

Two nearby concrete buildings offset this buoyancy: one contains toilets, and the other parking space for bicycles and motorbikes. A large patch of green on the waterfront, dotted with SANAA's own playful chrome-plated steel chairs, adds a touch of warmth to the unyielding coolness of the terminal's functional industrial aesthetic. The low-hanging terminal, however – like a piece of contemporary art – framed by thin columns and sharp roof lines, blends effortlessly with the harbour and sea.

OPPOSITE, TOP Contextual view. OPPOSITE, BOTTOM Polished walls act as structural stiffeners but remain visually ethereal. BELOW, LEFT The roof plane is supported with the minimum of help. BELOW, RIGHT The surrounding landscape is framed as well as reflected. BOTTOM LEFT A separate box houses security guards. BOTTOM RIGHT The playful public seating is a counterpoint to the sleek terminal.

INFRASTRUCTURE AND PUBLIC SPACES

Slowtecture M

Miki, Hyogo Prefecture

ENDO SHUHEI ARCHITECT INSTITUTE, 2003–2007

GROUND-FLOOR PLAN

BELOW, LEFT A grassy layer covers much of the building. BELOW, RIGHT The domed entrance resembles a giant tennis ball. OPPOSITE, TOP The sunken spectator court occupies the middle of the building. OPPOSITE, BOTTOM LEFT A triangular space frame supports the building's steel skin. OPPOSITE, BOTTOM RIGHT Clad with eye-popping yellow tiles, the foyer is hard to miss.

In January 1995 the Great Hanshin Earthquake shook Kobe so vigorously that entire neighbourhoods were levelled and portions of the regional infrastructure destroyed in a matter of minutes. Throwing the whole of Japan into a state of shock, the tragedy underscored the country's lack of preparedness for a natural disaster of such magnitude. In response, Hyogo Prefecture decided to turn a large, undeveloped tract of land 32 kilometres (20 miles) from the city into a disaster-relief complex that includes a fire-department training centre, an earthquake simulator and Shuhei Endo's striking Slowtecture M.

The winning entry in a proposal competition, Slowtecture M is an emergency shelter that doubles as a tennis dome. Resembling a giant bean partially covered with grass, Endo's organic form blends with the idyllic rolling hills near by, but is ready to spring into action in the event of another catastrophe. Since the building is enclosed with moveable glass panels at four perimeter locations, supply trucks can drive directly into the vast, column-free space, which can be quickly converted to accommodate displaced people. If the electricity fails, solar panels mounted on the roof can generate enough energy to light up the whole interior.

Fortunately, earthquakes registering high on the Richter scale are extremely rare, and so far players have had the run of Slowtecture M. Orientated towards its access road, the building's main entrance is shaped like an enormous tennis ball embedded in the earth, clad with yellow tiles and impossible to miss. The foyer opens on to the soaring interior with nine tennis courts: four on either side of a sunken spectator court. Draped with concrete rising up from the floor, the only enclosed spaces are the cafe and offices flanking the entrance and the locker rooms at the back.

The building is encapsulated in a thin skin of steel plates supported by a triangulated space frame on the inside and blanketed with a 60-centimetre-thick (2 ft) layer of soil and grass on the outside. Concentrated on the building's south side, where sunlight is the strongest, the grass padding steadily decreases on the shaded north side. The highly efficient structural system enabled the energy-conscious architect to save on steel at the outset, and the thick padding acts as insulation, helping to keep the indoor temperature comfortable all year round. Endo achieved this effect by planting ten different types of seed, irrigating them with piped rainwater. He expects wild grasses to take their place eventually, and the building's man-made materials to mellow. Although nature's destructive forces gave rise to Slowtecture M, its constructive energy will keep it going.

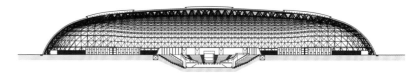

SECTIONS

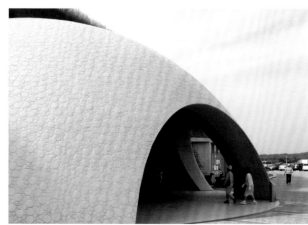

50 NEW ARCHITECTURE IN JAPAN

INFRASTRUCTURE AND PUBLIC SPACES 51

Tokyu Toyoko-Line Shibuya Station

Shibuya, Tokyo

TADAO ANDO ARCHITECT & ASSOCIATES,
TOKYU CORPORATION, NIKKEN SEKKEI AND
TOKYU ARCHITECTS AND ENGINEERS, 2006–2008

Tokyo undeniably has one of the most extensive and efficient mass-transit systems in the world. But the city's underground stations do not stack up: devoid of character and disconnected from the street, most are mazelike and disorientating. Shibuya Station, in one of Tokyo's main shopping and entertainment districts, is among the busiest and the most difficult to navigate. But when tracks for the Tokyo Metro's new Fukutoshin line were added to Shibuya's subterranean labyrinth, plans were also made for a destination station designed by Tadao Ando.

Usually, construction of the subway infrastructure precedes that of the station itself. This time, however, the two developed in tandem, creating a hub that will eventually unite the new line with the existing Tokyu Toyoko-Line. Conceived as an 'earth ship' floating under the ground, an ovoid volume nestling comfortably within the infrastructure's concrete frame, the idea recalls the Urban Egg Ando proposed for Nakanoshima, an island in his home town of Osaka (a project that was never built).

Inside the egg's hollow shell is a three-storey void that connects the station's below-grade concourse to the ticket gates below and the platforms at the very bottom. Spanning the various levels, massive columns tie the composition together, while escalators at either curved end move passengers up and down. Such an open, airy atrium not only relieves the congestion that plagues other stations, but also allows people to see where they are going.

A second but no less important asset of Ando's 'egg' is that it has yielded the world's first naturally ventilated underground station. Following the well-known principle that hot air rises, the atrium behaves like a giant chimney, drawing warm air from track level up and out. Ando selected lightweight, glass-reinforced concrete for the shell and laced it with water-filled pipes that passively chill the interior air.

Futuristic, but also reminiscent of Europe's elegant old train sheds, Ando's egg-shaped station sets a higher standard for train travel in Tokyo as it settles into its sunken location.

SECOND-FLOOR BASEMENT PLAN

FOURTH-FLOOR BASEMENT PLAN

OPPOSITE, BOTTOM LEFT Ticket gates at the station entrance. OPPOSITE, TOP A three-storey void connects the ticket gates, concourse and platforms. OPPOSITE, BOTTOM RIGHT View of train tracks. THIS PAGE An egg-shaped concrete shell encloses the new station.

INFRASTRUCTURE AND PUBLIC SPACES

Yokohama International Port Terminal

Yokohama, Kanagawa Prefecture

FOREIGN OFFICE ARCHITECTS, 1994–2002

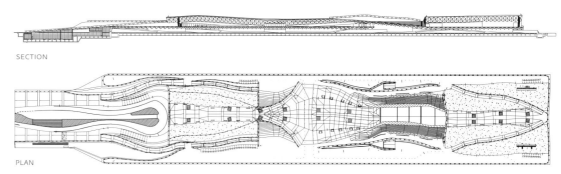

SECTION

PLAN

Yokohama is a city with a long history of cosmopolitanism. Aptly, then, the UK practice Foreign Office Architects won an international competition in 1994 to design its gateway to the world. The building is a relatively early demonstration of the power of computer-aided design (CAD): like the Guggenheim Bilbao (1997) or the roof of the British Museum's central court in London (2000), it would simply not have been possible without the advantages of CAD. Its complex forms, the integration of structure and surface, and the careful plotting of each piece of this three-dimensional jigsaw puzzle are the consequence of digital, rather than physical, modelling.

In fact, the word 'building' should be used loosely. Providing a terminal for domestic ferries and international cruise liners, the structure is more like a landform than anything else. Early in the design process, the architects imagined it performing a mediating role between land and sea, between building and park. The architects, who are known for their highly theoretical way of working, describe it as 'a public space that wraps around the terminal, neglecting its symbolic presence as a gate, de-codifying the rituals of travel, and a functional structure which becomes the mould of an a-typological public space, a landscape with no instructions for occupation'.

The structure emerges from the flat, linear cityscape, quickly becoming an undulating, warped space. Unlike a traditional pier, in which to exit one is merely required to retrace one's steps, this structure offers a variety of circulation routes and becomes a place for exploration. Built on 600 concrete-filled steel tubes buried deep in the seabed, the steel structure is composed of elements that merge floors with walls and provide broad, column-free interiors. However, the terminal's sculptural exterior provides little hint of what lies beneath: parking for cars and coaches; arrival and departure facilities; restaurants; meeting rooms and so on.

It is remarkable that so many function-driven spaces can be provided by an asymmetrical structure of slopes and blurred boundaries. This project cost ¥23 billion, provides 48,000 square metres (nearly 520,000 sq. ft) of space and caters for 53,000 people every year. Intriguingly, the timber used to clad the ground-level park – ipe, a Brazilian hardwood – is so dense that it does not float. With such a weighty presence, no wonder the terminal has become one of the most celebrated landmarks to emerge this side of the millennium.

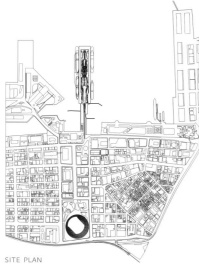

SITE PLAN

54 NEW ARCHITECTURE IN JAPAN

OPPOSITE, TOP More landform than building, the terminal mediates between the city and the sea. OPPOSITE, BOTTOM The landward approach to the terminal. ABOVE, LEFT Spaces are characterized by acute angles, faceted surfaces and blurred boundaries. ABOVE, RIGHT The terminal reinvents the pier, warping its traditional simplicity. FAR LEFT The crystalline ceiling treatment repeats the faceted language of the structure. LEFT The approach is imbued with a sense of drama.

INFRASTRUCTURE AND PUBLIC SPACES

Culture

58	21_21 Design Sight
60	21st Century Museum of Contemporary Art, Kanazawa
62	Aomori Museum of Art
64	Chichu Art Museum
66	Echigo-Matsunoyama Museum of Natural Science
68	Gallery Sora
70	Inujima Art Project: Seirensho
74	Kanno Museum of Art
76	Koshirakura Workshop Projects
78	Masanari Murai Memorial Museum of Art
80	Matsudai Snow-Land Agrarian Culture Centre
82	Nemunoki Museum of Art
84	Sendai Mediatheque
86	Suntory Museum of Art
88	Tomihiro Art Museum
90	Towada Art Center
92	Yokosuka Museum of Art

21_21 Design Sight

Minato, Tokyo

TADAO ANDO ARCHITECT & ASSOCIATES, 2004–2007

Tokyo, unlike London or New York, has no museum devoted to the history of design. But it does have 21_21 Design Sight, a cultural institution conceived by fashion icon Issey Miyake and architectural superstar Tadao Ando and dedicated to the future of the discipline. Draped with folded sheets of steel, its roof is a play on A-POC, Miyake's experimental clothing line in which each item is made from 'A Piece of Cloth', but the building's silky-smooth concrete walls bear Ando's distinctive signature.

The project began to take shape when the developer of the mixed-use Tokyo Midtown development offered the designers a site for their museum within the new green belt at the rear of the complex. The catch was that the

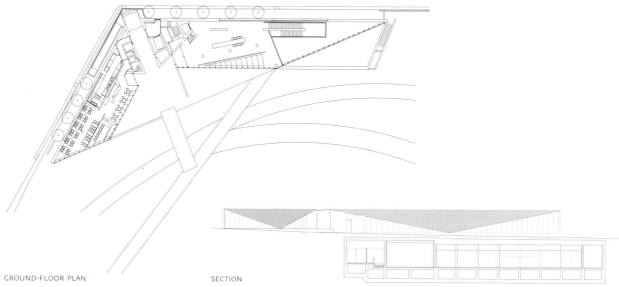

GROUND-FLOOR PLAN   SECTION

development's legal requirement for open parkland allowed them to build only 4.8 metres (16 ft) above ground. Yet there was plenty of room underground to create exhibition space for the site-specific installations commissioned by the curators. And because the museum does not maintain a permanent collection, it needs almost no storage space.

Marking the development's northwest corner, 21_21 Design Sight is a sleek triangular building consisting of twin trapezoids separated by an exterior passage. One volume contains the museum's single-storey cafe; the other its upper lobby, a generous space with Japan's only trapezoidal lift at one end and a floating concrete staircase at the other. Below are the lower lobby, curatorial offices and the two galleries: another trapezoid illuminated by indirect light from above, and a conventional rectangle with a ceiling an impressive 4.8 metres (16 ft) high.

Although buried below ground, the museum's main floor is remarkably light and airy thanks to a two-storey void filling the tip of the eastern building. Connecting top and bottom, the void is composed of two adjacent triangles, one a double-height interior space with a glass wall, the other a sunken courtyard open to the sky. The courtyard can be used as an entrance to the lower level for larger objects, but its primary role is to loosen up the corset-like concrete enclosure.

At street level, the museum's most striking feature is its remarkable steel roof. Figuratively anchored to the ground at the front by a round steel column, its seamless surface belies its complicated underpinnings. Despite its small size, 21_21 Design Sight's sculptural form makes a bold statement that stands up confidently to the towers of Tokyo Midtown.

OPPOSITE Aerial view. The steel roof is a play on a clothing line by the fashion designer Issey Miyake. ABOVE, RIGHT Floating stairs descend to the lower lobby. RIGHT The lower lobby abuts a sunken courtyard. FAR RIGHT Gallery interior.

21st Century Museum of Contemporary Art, Kanazawa

Kanazawa, Ishikawa Prefecture

KAZUYO SEJIMA + RYUE NISHIZAWA/SANAA, 2002–2004

An artful composition of white rectilinear volumes encircled by glass, the 21st Century Museum of Contemporary Art, Kanazawa, is so ethereal that it barely even looks like a building. It has no prominent façade and no primary entrance, and its thin skin practically melts away. Yet the minimalist museum has more presence than most public monuments made of concrete or stone.

The museum is the product of Tokyo firm SANAA (principals: Kazuyo Sejima and Ryue Nishizawa), which won the commission through a competition. The brief called for a combined museum and community centre to be constructed in the grounds of a former girls' school in the centre of town. Surrounded by streets and an ancient moat, SANAA's round building sits in a green lawn crossed with walkways that lead to its many entrances. All doors lead into the perimeter 'free zone', which is geared towards the general public and contains a restaurant, museum shop, rental gallery, art library, childcare centre and lecture hall. Reached by a glass-enclosed lift mounted on an enormous piston, and a grand staircase also surrounded by glass, the lower level holds the theatre, a second rental gallery and parking. The 'pay zone', controlled by ticket gates, occupies the core of the building, and is loosely separated from the free spaces by clear partitions, glassed-in courtyards and potted plants. Its fourteen independent galleries house the museum's permanent collection of post-1980 art, as well as temporary installations. Held together by an irregular grid of interior streets, the plan echoes that of the surrounding urban landscape.

While the building itself has a strong sculptural quality, the architects had no intention of competing with the art it was to contain. The gallery walls are painted white, the floors are of neutral concrete and the ceilings are made mostly of louvred glass to control daylight levels. Even the architect-designed furnishings, such as the white upholstered seating blocks, are recessive elements.

As well as creating spaces in which to display art, the architects had to incorporate six architecturally scaled works commissioned by the museum. They include James Turrell's *Blue Planet Sky*, a stone-encased room open to the heavens, and Leandro Erlich's remarkable crowd-pleaser *The Swimming Pool*, a shallow basin set over a chamber sunk into one of the museum's four courtyards.

In recent years many of Japan's smaller cities have built museums. Sadly, many have been labelled *hakomono* (empty boxes) because of under-utilization and weak curatorial leadership. Between its changing exhibitions and its community service, however, SANAA's 21st Century Museum of Contemporary Art, Kanazawa, seems poised for a rosy future.

ABOVE The round form contains both a community centre and an art museum. BELOW Aerial view. OPPOSITE, TOP LEFT The building's thin glass skin almost melts away. OPPOSITE, RIGHT Two views of Leandro Erlich's *The Swimming Pool*. OPPOSITE, BOTTOM LEFT One of the four internal courtyards.

Aomori Museum of Art

Aomori City, Aomori Prefecture

JUN AOKI & ASSOCIATES, 2000–2006

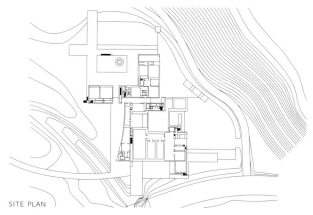

SITE PLAN

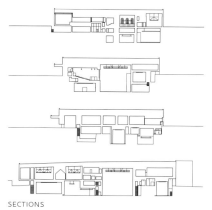

SECTIONS

Once a busy port, Aomori City sits snugly in a bay on the northern tip of Honshu, the main island of the Japanese archipelago. Its people are set apart by their distinctive dialect, and their quiet, introspective way of life often finds a powerful outlet in the visual arts.

In 2000 Tokyo-based architect Jun Aoki beat 392 other firms in a nationwide competition to design a major art museum for the city. The jury, which included Toyo Ito, was impressed with Aoki's idea of submerging the structure underground, a concept that enabled the museum to 'reflect' the excavation process going on next door at Sannai Maruyama, one of Japan's largest and oldest Jomon period (8000–400 BC) archaeological sites. Aoki made the reference by cutting the ground beneath the museum into trenches, both inside and out, some of them 20 metres (66 ft) deep.

The excavated earth, mixed with cement, was used as the base and capped by a white steel structure – an allusion perhaps to the snow-capped mountains of Aomori – containing cuboid galleries. Countering its subterranean associations, the dramatic interlocking of the pounded brown earth (assimilating a traditional building method called *tataki*) and white-plastered stucco finishes inside the museum is anything but dingy. Daylight streams into the giant spaces to stunning effect.

The complex is entered either via lifts from the main gate or down a ramp, which is reserved for visitors coming from or going to the Sannai Maruyama site. Workshops, a cafe, a library and office spaces are all neatly contained in the upper level of the white volume, while the building's main circulation elements – corridors, entrances and courtyards – have been fitted within the interstitial spaces.

In sharp contrast to the bright galleries is Aleko Hall, the museum's main exhibition room, which is darkened to display Marc Chagall's paintings, once the backdrops for the ballet *Aleko*. Aoki adds a theatrical flourish with a moveable wall that reveals the auditorium flanking the room, so that the ballet may be performed appropriately, the paintings appearing almost as mirages in the distance.

Despite its typically Modernist arrangement of white cubes, the museum appears in many ways as old-fashioned as a nineteenth-century collection of curiosities: full of secrets and intrigue. In this sense, it seems wilfully to reject the current vogue for transparency and openness. Ornamental details, such as the partially visible steel frame, the curtains draped over the windows and the white-painted brick curtain walls, heighten the suspense, enriching the labyrinthine network of chambers and avenues.

ABOVE AND OPPOSITE, TOP LEFT The main entrance to the museum, and its sculptural approach. OPPOSITE, TOP RIGHT The Octagonal Chamber was a collaboration between artist Yoshitomo Nara and architect Jun Aoki. OPPOSITE, BOTTOM LEFT An exit leads from the basement reception directly to the neighbouring archaeological site, Sannai Maruyama. OPPOSITE, BOTTOM RIGHT A courtyard in one of the 'trenches' houses Nara's 8.5-metre-tall (27⅞ ft) *Aomori-ken Dog*.

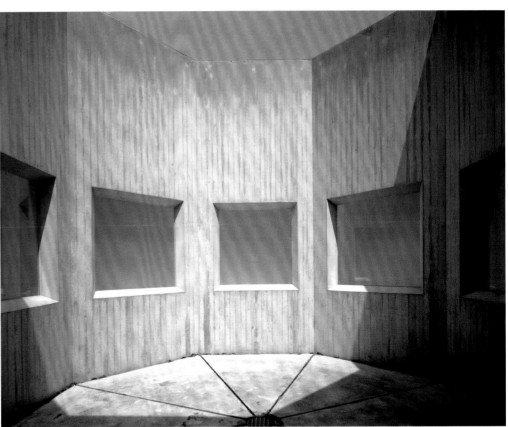

CULTURE 63

Chichu Art Museum

Naoshima, Kagawa Prefecture

TADAO ANDO ARCHITECT & ASSOCIATES, 2000–2004

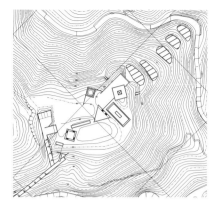

SITE PLAN

BELOW Aerial view. OPPOSITE, CLOCKWISE FROM TOP LEFT: Tilting concrete planes enclose an exterior walkway; two exterior courtyard details; *Time/Timeless/No Time* by Walter De Maria; a rest area with benches.

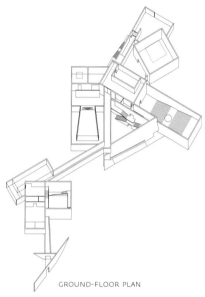

GROUND-FLOOR PLAN

The Chichu Art Museum is a monumental work of architecture minus the monumentality: a cluster of bold geometric forms rendered in the architect's favourite material, concrete, but embedded in the ground (*chichu* means 'within the earth'). Instead of standing out, they sit comfortably in the hilly landscape of Naoshima, a small island in the Seto Inland Sea.

Accessible only by boat or ferry, Naoshima, like several of its neighbours, developed as an industrial site at the end of the nineteenth century. Since the decline of industry on the island, the private Naoshima Fukutake Art Museum Foundation has been gradually acquiring pieces of land and transforming it into a cultural Mecca. Its first foray was Benesse House of 1992 (also designed by Ando), a combined hotel and gallery. The nearby Chichu Art Museum features permanent installations of works by three artists – Claude Monet, Walter De Maria and James Turrell – each in its own self-contained space.

A concrete wall slicing across the hillside marks the edge of the museum precinct and the start of the labyrinthine sequence of spaces that precedes the galleries. Distancing the museum's refined inner sanctum from its untamed surroundings, a tunnel-like passage leads first to a square forecourt carpeted with straight stalks of scouring rush. As in many Japanese gardens, the lush greenery is for the eyes only, but a perimeter path provides many vantage points before depositing visitors in the lobby.

Open to the sky but confined by tilting concrete planes, a second walkway leads from the lobby to the heart of the museum. This triangular void has a limestone-strewn base marking the museum's lowest point, but its sheer concrete walls also enclose the tallest (albeit sunken) space. From there, exterior stairs descend to the De Maria gallery and a concealed ramp ascends to the Turrell and Monet galleries. The circuitous route culminates in the cafe, where a floor-to-ceiling window wall opens to the sea, reuniting Ando's tightly controlled interior with its unbounded natural environment.

Ando provided a shell for each exhibit, but the installations were the purview of the living artists and the Chichu's curator; as a result, the experience in each gallery is unique. But all the works, in their exploration of the theme of light, are consistent with Ando's architecture.

Because of the museum's waterside location, the ravages of salt-laden sea air and high humidity were serious considerations where the artworks were concerned. However, Ando hopes that the forces of nature will gradually erode the museum's exposed surfaces, causing his architecture to recede even further from view.

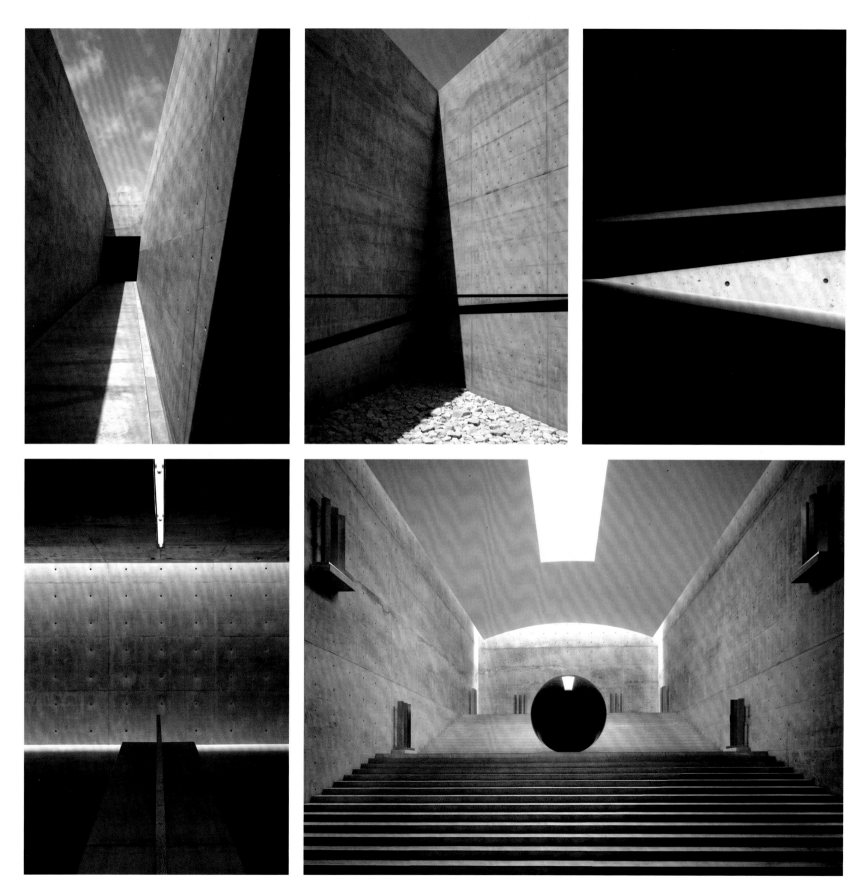

Echigo-Matsunoyama Museum of Natural Science

Matsunoyama, Niigata Prefecture

TEZUKA ARCHITECTS, 2002–2003

SITE PLAN

GROUND-FLOOR PLAN, WITH TOWER

Built like a submarine, the Echigo-Matsunoyama Museum of Natural Science is a hollow tube of Cor-ten steel that crawls along the ground but terminates dramatically in a lookout tower resembling a giant periscope. Encased inside, 160 steps ascend in semi-darkness, rewarding the intrepid with spectacular panoramic views of a snow-covered landscape in winter and a patchwork of rice paddies in summer.

Located in the agricultural heart of Niigata Prefecture, the museum is one of several public works to be built in conjunction with the Echigo-Tsumari Art Triennal. Tokyo-based husband-and-wife team Tezuka Architects was awarded the job after a design competition, with a plan that traces an abandoned path that once linked the terraced rice fields.

While naturally focusing on the museum's exhibits – which feature local flora and fauna as well as a donated butterfly collection – the architects also wanted to create a place that would encourage visitors to observe and reflect on the surrounding environment. They have achieved both with a linear sequence of spaces punctuated with picture windows where the building bends. From the lobby, a left turn leads to the galleries and then the tower, and a right turn to the offices, a laboratory, the function hall and, finally, the cafe. Research facilities are squirrelled away on the first floor.

Because of the region's unusually high snowfall – an astounding 30 metres (98 ft) falls annually – the architects had to forgo conventional construction. Made of metal plates welded on site, the museum's double-layered steel shell required the expertise of shipbuilders. The large display windows that frame

OPPOSITE, TOP The industrial aesthetic of the museum stands out in the snow. OPPOSITE, BOTTOM The approach to the main entrance. LEFT Rear elevation. BELOW, LEFT An aquarium-grade acrylic window frames the cross-section of a snow drift. BELOW, RIGHT The museum's butterfly collection.

cross-section views of compacted snow are made of aquarium-grade acrylic, since ordinary glass would impart a greenish cast to the white drifts that blanket the ground from December to April.

While the building's bold form stands out from the site, the architects hope the museum will blend comfortably with its natural environment. As its industrial materials gradually take on the reddish hue of the site's iron-rich soil, and indigenous plants take root around the building, it seems to be heading in the right direction.

CULTURE 67

Gallery Sora

Chuo, Tokyo

AKIHISA HIRATA ARCHITECTURE OFFICE, 2007–

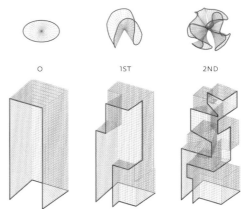

The idea of architecture without a floor seems impossible. But if Gallery Sora is realized, Tokyo-based architect Akihisa Hirata might just achieve this lofty goal. A direct commission from an established gallery owner, the project will replace a conventional building used for art storage with a cascade of interconnected spaces held together by a single pleated surface that riddles the building.

Because the client knew exactly what he wanted – a tall exhibition space for contemporary art at street level, a cafe on the first floor, and two apartments, one each on the second and third floors – the building's vertical organization was prescribed. But the architect had no intention of fulfilling these needs with standard slab-and-column space. On the contrary, he envisioned an organic composition inspired by a plume of smoke.

With that in mind, Hirata embarked on a rigorous form-finding mission, beginning with a small cubic study model with dimensions derived from the site, a corner plot near the high-end galleries of Tokyo's Ginza commercial district. His first move was to outline the six edges where the model met the ground and faced the streets. He then 'pleated' this single continuous line, turning its straightforward profile into a convoluted one. Where the line indented, Hirata pared away the volume to articulate the box. Where the line remained unchanged, it retained the model's rectangular mass. Hirata then repeated these steps, folding the already folded line and excising more solid matter, until he reached the right composition: a vertical chain of semi-enclosed spaces defined by an unbroken wall representing the line.

Only after he had determined the intricacies of the building did Hirata address its programme. Changing the abstract form into usable space entailed a study of possible room adjacencies and a creative reinterpretation of the client's requests: the cafe, for example, is indeed above the main gallery, but is divided between two levels. Hirata would not compromise the irregular geometry of the individual spaces nor detract from his pleated walls by adding partitions or other intrusive structural elements. Instead of columns and beams, the support system is integrated with the architectural components. Hirata's only nod to practicality was to insert sheets of glass to enclose the rooms and staircases to connect them.

Officially, Gallery Sora is a six-storey building, but the architect happily admits he lost count after the first floor, where short runs of stairs link myriad staggered levels. Yet he never loses sight of the line. Rendered with rough concrete on one side and pure white paint on the other, it remains the dominant feature of a building that, when finished, will be as much a work of art as a piece of architecture.

RIGHT, TOP Conceptual diagrams. RIGHT Axonometric. OPPOSITE, LEFT Three views of the galleries. OPPOSITE, RIGHT A computer-generated image reveals the gallery's sequence of cascading spaces.

Inujima Art Project: Seirensho

Inujima, Okayama Prefecture

SAMBUICHI ARCHITECTS AND ARUP JAPAN, 2006–2008

In 2007 Japan's Ministry of Economy, Trade and Industry finally moved to recognize various disused pre-war industrial sites as potential catalysts for regional redevelopments. Thirty-three industrial heritage sites across Japan were duly chosen to be preserved and used. On the list was Inujima Copper Refinery, built in 1909, which had closed down within ten years of opening.

The small island of Inujima is a mere ten minutes' ferry ride from the mainland of Honshu on the Seto Inland Sea. Soichiro Fukutake, the president of Benesse Corporation, acquired the site of the refinery in the late 1990s. In 2006 he commissioned Hiroshi Sambuichi, a young architect renowned for his ecological housing schemes, to design an ambitious art project that would put the island on the architectural map.

The architect – whose office is in Hiroshima, where he grew up, and who is therefore familiar with the history behind many of the islands on the Seto Inland Sea – set out to reuse part of the ruin on Inujima and to preserve the rest. Collaborating closely with the internationally acclaimed artist Yukinori Yanagi, Sambuichi has created a low-rise, timber-clad, semi-subterranean museum using locally available materials, such as the lime-coloured granite that has been quarried on the island for centuries, and dark-grey *karami* bricks, a factory by-product once dumped in the sea. Sambuichi, who considers landscaping an important part of his job, has also managed to dispense altogether with the need for artificial light and ventilation, rendering the museum more or less an extension of the landscape.

Visitors on guided tours are first introduced to the Earth Gallery, a dark, cave-like subterranean corridor, with

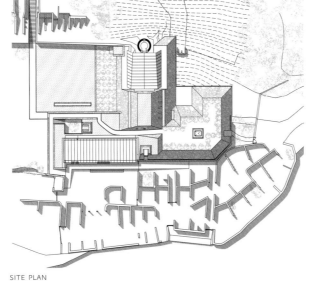

SITE PLAN

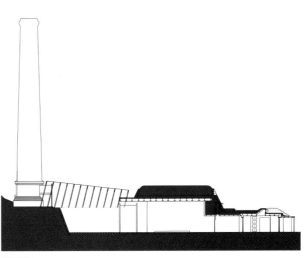

SECTION

OPPOSITE The original chimney naturally ventilates the museum. LEFT *Karami* bricks, a factory by-product once dumped in the sea, have been gathered to sculpt the approaches to the gallery and the rest area near the sea. BELOW Hot and cold air converge in the Energy Hall. BOTTOM LEFT The Chimney Hall contains floating bits and pieces from the house of the famous writer Yukio Mishima. BOTTOM RIGHT The corrugated steel walls of the Earth Gallery maximize the conduction of cool subterranean air. OVERLEAF Warm air collects in the glazed Sun Gallery.

a wavy Cor-ten-steel wall and long winding path that circulate the cool air. The main gallery is a large, arched shelter called the Energy Hall. Two large sliding doors can be closed or opened to bring in either chilled air from the Earth Gallery or warmed air from the Sun Gallery, a conservatory containing another exhibition space and a cafe.

The floating exhibits are pieces taken from an old timber house that once belonged to Yukio Mishima, the famous Japanese writer who committed ritual suicide inside an army headquarters in 1970. Fukutake was asked to purchase the lot when it was decided that the house must be disposed of. Sambuichi's subtle integration of Japan's industrial past with its sustainable future has been made perceptible, therefore, through another ruin, more intimate and personal.

Kanno Museum of Art

Shiogama, Miyagi Prefecture

ATELIER HITOSHI ABE, 2003–2005

The Kanno Museum of Art on the outskirts of Sendai is a pure metal box built to house a private collection of metal sculptures. The cube of Cor-ten steel belongs to a psychiatrist with a passion for figurative works. Eager to open her collection to the public, she acquired a large parcel of land next to her house and entrusted Hitoshi Abe with the realization of her dream.

Faced with this challenge, Abe began by drawing a bubble around each of the eight sculptures that form the core of the collection. He then imagined piling the spheres into a box that would hold them in place and create points of contact. As the project progressed, the one-to-one correlation between the artworks and their individual enclosures weakened, since he had to plan for temporary exhibitions and a small concert space as well. But this turned the bubbles into a bona fide building.

Naturally, this organic arrangement was not suited to conventional construction methods. In collaboration with a structural engineer and a shipbuilder well versed in the vagaries of steel-plate construction, Abe invented a pioneering wall system consisting of two steel sheets embossed with a grid of oblong indentations, welded back-to-back. The interior surfaces were coated with white paint or plasterboard for hanging art, while the exterior surfaces have taken on a rusty patina.

The museum comports itself on its bed of green grass like a sculpture meant to be admired from all sides. Although the doctor has direct access to the building from her home next door, visitors approach the hillside site by a narrow road that winds its way between small rice paddies and modest homes – a common sight in many Japanese suburbs. A short run of stairs leads to the museum entrance, which is set away from the street.

Inside, Abe's architecture has a dizzying effect, where ceilings become walls and walls become floors. Leaving its surroundings behind, the museum begins with a reception area, from where a short run of stairs leads down to the first in a sequence of galleries dispersed over three descending levels. The sequence culminates in a large room suited equally to art or to music. Completing the circulation loop, a lift leads back up to the reception area, with its small triangular window looking out towards the distant sea. Elsewhere, an occasional slash-shaped opening lets in a modicum of light and forges a connection to the immediate environment. But the surroundings are of little importance to a museum that provides the chance to lose oneself in a world focused solely on art.

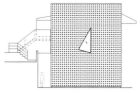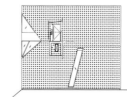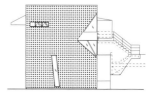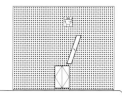

ELEVATIONS

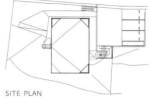

SITE PLAN

ABOVE Concrete stairs lead up to the main entrance. OPPOSITE, CLOCKWISE FROM TOP LEFT A window detail; designed for eight sculptures, the main gallery fills the bottom of the building; stairs descend from the reception area; concrete stairs contrast with the museum's steel exterior.

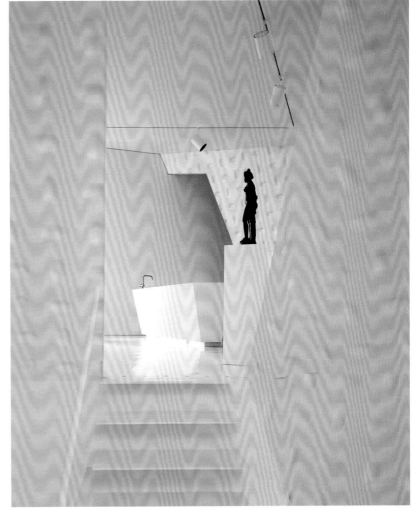

CULTURE 75

Koshirakura Workshop Projects

Koshirakura, Niigata Prefecture

SHIN EGASHIRA AND STUDENTS FROM
THE ARCHITECTURAL ASSOCIATION
SCHOOL OF ARCHITECTURE, 1998–

BELOW Farmhouse project with a viewing platform that extends from within the house. BOTTOM LEFT Sketch of the intervention. BOTTOM RIGHT Interior view of the ongoing work at the farmhouse.

The small village of Koshirakura, 200 kilometres (124 miles) north of Tokyo, is one of many in rural Japan that are experiencing acute demographic decline owing to the opportunities offered by the country's vibrant and wealthy cities. But since 1998 the London-based architect and lecturer Shin Egashira, who teaches at the prestigious Architectural Association School of Architecture, has organized annual workshops in which students work alongside elderly villagers and create experimental buildings. Mostly self-funded, partly supported by the local government, and using the village's disused primary school as a base, the workshops have provided Koshirakura with a bus shelter, artworks and even an open-air cinema screen constructed from parallel strands of taut rope. Students and locals collaborate to undertake all the projects during the summer.

One of the more interesting and arresting outcomes of this long association between village and architecture school is the creation of a flexible summer pavilion, *Azumaya*, which explores the idea of the traditional garden hut. Built as shelters from sun and rain, these little structures have, in Japan, always adhered to certain rules: they tend to be asymmetrical and have no walls, and they are not supported on four columns (apparently to avoid connotations of the space underneath a four-legged animal). The Koshirakura pavilion was designed by Egashira and students in the winter of 1998/99; a local contractor offered to teach students time-honoured woodworking techniques and guaranteed timely completion.

The finished building is a two-storey structure that can be closed in the winter, with a roof that sheds snow. During the summer, the building opens up so completely that its roof appears to be suspended lightly above the ground. Two covered bench systems can be pushed out of the building along tracks, allowing the structure to reach out into, and frame, the landscape. A number of configurations are possible to accommodate events of different types, including karaoke parties and tea ceremonies. Vertical screens flip upwards to become horizontal shading devices, and upper-level louvres can be adjusted to moderate the sunlight.

More recently Shin Egashira and his student collaborators have focused their attention on an existing farmhouse, which has been renovated and fitted with a mechanism by which an elevated viewing platform can be slid out on to a pier. This project is, perhaps, a quirkier intervention than *Azumaya*, but all the work undertaken during these annual workshops could be described as unorthodox. The beguiling designs are deliberate attempts to tap into local memory and perceptions of landscape, and their artistic inventiveness is just as important as their function.

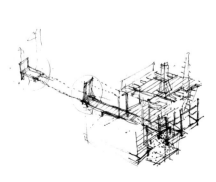

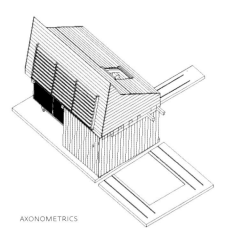
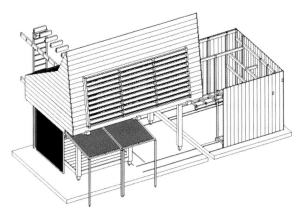

AXONOMETRICS

TOP LEFT *Azumaya* in open, summer mode. RIGHT A series of photographs shows the gradual extension of the building along tracks. BOTTOM LEFT The asymmetrical roof is designed to shed snow.

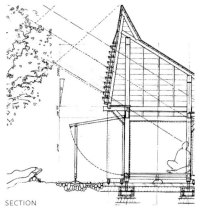

SECTION

CULTURE 77

Masanari Murai Memorial Museum of Art

Setagaya, Tokyo

KENGO KUMA & ASSOCIATES, 2001–2004

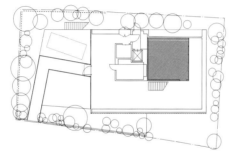
SITE PLAN

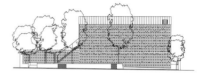
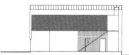

ELEVATIONS

Masanari Murai Memorial Museum of Art showcases not only the artist's work but also the studio in which the artwork was produced. This is not the first time that an artist's studio has been preserved permanently inside a museum: the South Kensington studio of Murai's contemporary Francis Bacon, and all the ephemera contained within it, was relocated to Dublin City Gallery The Hugh Lane in 1998.

Like Bacon, but less maniacal, Murai hoarded everyday objects, and whenever he ran out of space he simply extended his studio rather than part with any of his collection. A maze consequently evolved, with corridors that led nowhere and rooms that could be accessed only by climbing through internal windows. Inhabiting these labyrinthine channels, Murai lived, worked and taught painting until his death in 1999 at the age of ninety-three.

Captivated by this extraordinary space, Kuma decided to incorporate the studio's various components into his design. As well as the studio itself, the original exterior timbers, complete with holes and scratches, have been saved and reassembled as louvres on the façade. Tall native trees shade the reinforced-steel building, and a concrete walkway cuts through raised Cor-ten-steel pools. At night, the museum seems to rise out of a much deeper pond. To complete his artistic excavation, Kuma salvaged the painter's beloved Toyota Crown car after twenty years of disuse, and displayed it as an outdoor sculpture sitting in the water in front of the museum.

The original studio is enclosed within an L-shaped exhibition hall, stretching across the total floor space of 300 square metres (3230 sq. ft). Its walls – patched up in parts – reveal the

OPPOSITE, TOP The museum is approached past raised Cor-ten-steel pools. OPPOSITE, BOTTOM The artist's beloved Toyota is preserved as a piece of outdoor sculpture, and the reception area is decorated with furniture from his old house. BELOW Museum interior. RIGHT AND BOTTOM The original walls, doorframes and windows of the artist's studio, as well as all the objects inside it, have been faithfully restored inside the new museum.

jigsaw that was the artist's house, and the fervour that must have consumed him. A wafer-thin steel staircase permanently blocks the original way into the studio and, instead, leads visitors to a small landing, a transitional space between public and private, at the back of which is the artist's widow's residence. This new home now sits protectively over what was so close to the artist's heart.

Despite the unusual totem-like pole at the gate, subtly declaring it to be a public building, the museum has an intensely private feel: 'It is still a house, even though it's located inside an art museum,' Kuma says, carefully emphasizing the multi-purpose aspect of the museum. 'I wanted to maintain this sense of duality, as I thought that the tenderness of a home is well suited to Murai's art.'

CULTURE 79

Matsudai Snow-Land Agrarian Culture Centre

Matsudai, Niigata Prefecture

MVRDV AND SUPER-OS, 2000–2003

Located in the mountains to the west of Japan, Matsudai experiences hot, humid summers and cold, snowy winters. Snow can lie in drifts up to 3 metres (10 ft) deep. This building, which contains art space and community facilities, is raised above the ground to provide a shaded plaza in the summer months and to be high above the snow during the winter. In 2000 the Rotterdam-based practice MVRDV, together with Tokyo firm Super-OS, took on the commission by the municipality of Matsudai and Tokyo's Art Front Gallery to design the centre. It is MVRDV's first completed project in Japan.

Theoretically, the centre celebrates the potential for art within the broad themes of agriculture and snow, but – more importantly perhaps – it also functions as the principal venue for the Echigo-Tsumari Art Triennial, a festival that aims to rejuvenate this rural area, with its ageing population. In the past, not only artists but also architects, such as Atelier Bow-Wow, have been called on to create temporary and sometimes permanent art installations across the region.

Located not far from the main road, between a river and a railway line, the centre is a steel structure, its square volume lifted off the ground by a series of covered stairways positioned against the façades at acute angles. The structure of the building is integral to its space planning and interior design. One of the spidery 'legs' stretches out to the walkway that leads to the railway station. The legs also pierce the building right through, forming corridors and creating six triangular and trapezoidal spaces in between. Five of these spaces (a multi-purpose room, a classroom, a cafe, an office and toilets) are positioned on the perimeter of the

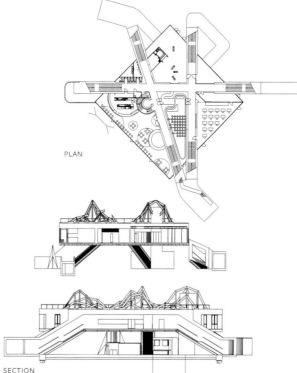

building; the wedge-shaped shop is towards the centre of the building, enclosed by the converging paths of two of the legs.

The form and colour of the building, although not immediately contextual, are not entirely alien to its locale. Distinctively white against surrounding greenery in warmer periods, the building all but disappears when the snow arrives; and the roof's asymmetrical steelwork, which pushes up from below in clusters of criss-crossing beams, is an echo of the surrounding hills. The intentionally bright colour schemes and lighting within the building (some of them arrived at through collaborations with artists) echo the intense inwardness of snow-trapped houses.

OPPOSITE, TOP The centre rises above the snowy landscape. OPPOSITE, CENTRE The building's 'legs' contain stairs. OPPOSITE, BOTTOM The roof structure is complex. RIGHT Views from the cafe run the length of the façade. BELOW, LEFT The main exhibition space. BELOW, RIGHT *Blackboard Classroom* by artist Tatsuo Kawaguchi.

Nemunoki Museum of Art

Kakegawa, Shizuoka Prefecture

TERUNOBU FUJIMORI, 2004–2006

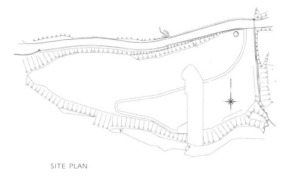
SITE PLAN

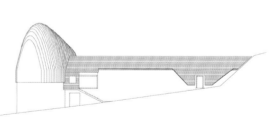
ELEVATION

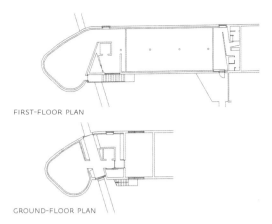

FIRST-FLOOR PLAN

GROUND-FLOOR PLAN

Nemunoki Museum of Art is set at the end of a private road, past a cluster of buildings that belongs to Nemunoki Gakuen, an institute for physically disabled children set up by the celebrated Japanese singer and actress Mariko Miyagi in 1967. To mark the institute's fortieth anniversary, Miyagi commissioned Terunobu Fujimori to build a new museum to showcase Nemunoki Gakuen's extensive collection of children's artwork.

Jutting out from the side of a hill, the museum sits in an idyllic valley of lush green tea fields surrounded by a thick forest of pine, cherry, cedar, chestnut and persimmon trees. A line of living grass on the ridge of the oblong structure, with its top-heavy dome covered in hand-rolled copper sheets, connects the museum to the hill behind it. Fujimori borrowed the idea from *shibamune*, a traditional method that encourages plants to take root in the layer of thatch. In this case, however, a bed of soil was built into the rooftop to keep the grass alive. It humorously completes the architect's imagery of 'a hairy mammoth about to stir'.

Carefully choreographed by Fujimori, the surrounding environment plays a part in the scheme, as visitors are led into the museum through the reception area under the dome and immediately out again to follow a circular path around the landscaped garden, before re-entering the building through a small, cave-like entrance at the back. The detour works like a purification ritual, creating a period of reflection, and the small door (acting in much the same way as the diminutive entrance of a traditional tea house) gently reminds visitors to be respectful of the work they are about to see.

In contrast to the dynamic exterior contour, the museum's interior is simple and modest. The children's colourful paintings and intricate drawings line the white plaster walls. A bold, free-standing wooden column chiselled by the architect himself from a tree trunk separates the first and second exhibition rooms. With the timber much used elsewhere in the museum, it adds an appropriate warmth.

A surprise is in store towards the end of the tour. The domed ceiling of the womb-like, white-painted second exhibition room has a long, narrow skylight covered with wooden latticework that tapers at one end, recalling the mammoth, the 'skeleton' of which can now be seen. The enfolding space is enchanting. The architect's whimsical reference to prehistory dispels any clinical coldness typically associated with modern white-cube galleries. Meanwhile, truly moving pieces of artwork are gently left to dominate.

OPPOSITE The dome, or 'hump' of a mammoth, emerges from the mist. TOP The exhibition spaces are simple but effective. ABOVE, LEFT Another exhibition room under the dome's 'spine'. ABOVE, RIGHT A timber column, shaped by the architect from the trunk of a chestnut tree using a barking machine, divides the exhibition rooms.

Sendai Mediatheque

Sendai, Miyagi Prefecture

TOYO ITO & ASSOCIATES, ARCHITECTS, 1995–2002

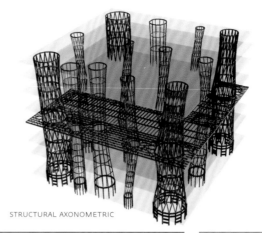

STRUCTURAL AXONOMETRIC

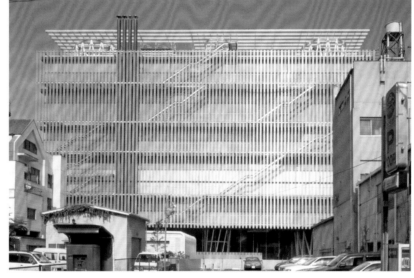

'As audacious as the Pompidou Centre', is how Michael Webb, writing in *Architectural Review*, described Toyo Ito's Sendai Mediatheque. With more than 1 million residents, Sendai is a major regional city; unusually for Japan, it is also a well-planned one, with plenty of broad green avenues tidily branching out from its main railway station. While the city has always attracted tourists, Sendai Mediatheque placed it firmly on the map of 'cool' cities.

Ito won the commission in 1995 through a competition, the judges for which included Arata Isozaki. Subsequently, however, it took nearly six years of debate to convince various groups, including librarians and curators, of the wisdom of his radical idea: a space that was fluid, transparent and open, where diverse activities within the building would be revealed to people strolling outside. Retrospectively, it seems incomprehensible that the design presented such a problem, considering the wide-ranging support it now has from the architectural establishment. SANAA, for example, has been spearheading the 'open-plan' trend in public buildings, and the younger generation of architects continues to experiment, pushing the concept to its limits.

Ito and structural engineer Mutsuro Sasaki managed to convince the people of Sendai to let them build a structure that almost eliminated the façade, or 'skin' as the architect liked to refer to it at that time. A series of columns – each one an assembly of white steel tubes either 160 or 240 millimetres (6⅜ or 9½ in.) in diameter – penetrates the floors to bear the load of the building and draw in natural light. Freed from supporting the structure, the exterior walls are clad in clear glass etched with

SIXTH-FLOOR PLAN

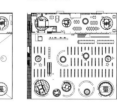

ROOF PLAN

SECOND-FLOOR PLAN

THIRD-FLOOR PLAN

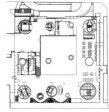

GROUND-FLOOR PLAN

FIRST-FLOOR PLAN

horizontal bands of abstract decorations to complement the remarkably thin floor slabs of steel and concrete, visible from the street. Ito likens the columns to strands of seaweed waving inside an aquarium, but the thinner columns show more of the arboreal vitality of the tall zelkova trees lining the avenue in front of the museum.

Four of these columns contain lifts, allowing visitors to see through all seven floors at once. Each floor possesses a flavour markedly different from the others, as Ito gave such young designers as Kazuyo Sejima, Karim Rashid and Ross Lovegrove free rein over 20,000 square metres (215,000 sq. ft) of floor space, including the fixtures and furnishings. Most recently, Osamu Tsukihashi of Architects Teehouse was given the opportunity to create an experimental cafe on the top floor, equipped with threaded-steel tables and chairs and ceramic floor tiles, to celebrate the opening of the fifth column, which contains stairs rather than a lift. Thanks to its fluid nature, Sendai Mediatheque continues to evolve.

OPPOSITE, TOP LEFT Back elevation. OPPOSITE, TOP RIGHT The experimental cafe by Architects Teehouse. OPPOSITE, BOTTOM The spacious reception area on the ground floor. RIGHT Front elevation, showing some of the thirteen tubular steel columns that penetrate through seven steel-ribbed floor slabs to bear the load.

Suntory Museum of Art

Minato, Tokyo

KENGO KUMA & ASSOCIATES, 2004–2007

A respite from the city, the Suntory Museum of Art was envisioned as a Japanese-style room but is embedded in the Tokyo Midtown development, a monumental commercial complex made of glass and steel. The museum, which belongs to a large brewery and distillery, contains one of the country's finest private collections of traditional art and historic objects. Fittingly, its new home is designed by Kengo Kuma, a Tokyo-based architect who champions local materials and the craft of carpentry.

Clad in Kuma's signature element, a delicate screen rendered here with ceramic fins, the museum has neither exterior signage nor direct street access. A cubic volume that reads from the outside as an independent entity, it occupies portions of Tokyo Midtown's third, fourth, fifth and sixth floors; visitors enter from the third floor of the multi-storey shopping centre. The museum's corner entrance leads to the shop and cafe in one direction and internal lifts in the other. Self-contained and sequestered from Midtown's commercial corridors, the exhibits begin on the fourth floor, where the main sequence of display spaces unfolds. Elegant glass-enclosed stairs and a double-height atrium connect to an additional gallery on the third floor. While the fifth floor holds museum offices and storage, the sixth overlooks the city with a terrace-wrapped conference hall and a private club as well as a tea-ceremony room – Kuma's take on one of traditional Japan's most highly ritualized spaces.

A mixture of old and new components, the tea room reincarnates a woven cedar ceiling, sliding paper partitions and other elements from the museum's previous quarters near by. Although Kuma had to work within the

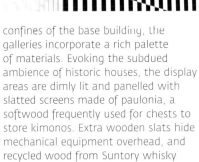
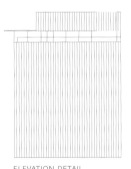
ELEVATION DETAIL

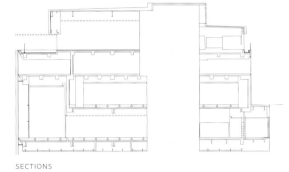
SECTIONS

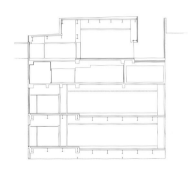

confines of the base building, the galleries incorporate a rich palette of materials. Evoking the subdued ambience of historic houses, the display areas are dimly lit and panelled with slatted screens made of paulonia, a softwood frequently used for chests to store kimonos. Extra wooden slats hide mechanical equipment overhead, and recycled wood from Suntory whisky barrels covers the floors.

At the perimeter of the museum, still more slatted wood screens filter contact with the internal shopping concourse and the external landscape. Inspired by *musogoshi* shutters, which are traditionally used to control light and views, the screens are formed of overlapping panels that can be pushed aside to open the atrium to the park at the back of the new development.

Striking the perfect balance between exposure and enclosure, Kuma's myriad screening devices gently filter out the surroundings and focus attention where it belongs: on the museum's magnificent paintings, textiles and ceramics.

OPPOSITE The museum is embedded in the mixed-use Tokyo Midtown development. ABOVE Detail of the exterior. TOP LEFT Glass-encased stairs rise through the double-height gallery. TOP RIGHT The tea room. RIGHT The galleries.

CULTURE 87

Tomihiro Art Museum
Azuma, Gunma Prefecture

AAT AND MAKOTO YOKOMIZO, ARCHITECTS, 2003–2005

In 2003 Makoto Yokomizo won an international competition organized by the village of Azuma to design a new art museum to showcase the work of the hugely popular local artist Tomihiro Hoshino. Hoshino, who is paralysed from the neck down, uses his mouth to paint vibrant and beautiful images of flowers and plants and to write poignant poetry.

The museum's unusual internal structure can be best observed from a nearby hill, from where one can see how cylindrical rooms of varying sizes are squeezed together inside a large rectilinear box. This is an almost complete compositional reversal of SANAA's 21st Century Museum of Contemporary Art in Kanazawa (completed in 2004; see pages 60–61), where cubes are enclosed within a circular perimeter. It is interesting to note that Kazuyo Sejima of SANAA and Yokomizo both worked for Toyo Ito before they set up their own offices, although Ito's characteristic transparency and lightness seem to have had less effect on the work of Yokomizo than on that of Sejima.

Further observation from above reveals that the circles at the edges of the square plan are brutally cut off to fit; otherwise the circular shapes are retained, with the interstitial spaces left open to the elements. The self-supporting steel cylinders, which were produced by a silo manufacturer, are as thin as possible (the joints where two walls meet are just 38 millimetres/1½ in. thick). In varying shades of grey and all at the same height, they resemble the ripples of water on the nearby lake, which, despite being man-made, is spectacularly beautiful and the dominant feature of the surrounding landscape.

Yokomizo treats the cylinders as if each exists in its own bubble. Visitors experience a random sequence of these 'bubbles', which are made freely accessible by being linked and opened up in as many ways as possible. The large central lobby and some exhibition rooms shooting off it are introspective, cut off from natural light. Others are punctured with low-level portholes, allowing some light to enter from the courtyards between the cylinders. The museum's cafe, shop, office, storage spaces and services are all placed around these exhibition spaces, along the edges of the building, where plenty of natural light streams in. The rooms are colour-coded and each is harmoniously and individually finished with a range of materials.

The key to the success of this museum is perhaps the fact that the young architect's idealism seeps through its programming. It is a friendly place, offering unlimited and equal access to people whose movements may be restricted in one way or another, allowing them to experience it in a celebratory and democratic way.

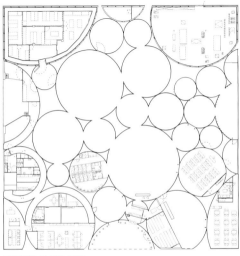

OPPOSITE Aerial view. TOP LEFT The museum cafe is placed in a corner to maximize the excellent view of Lake Kusaki. ABOVE The reflective ceiling not only frames but also magnifies the surrounding landscape. CENTRE LEFT Some of the interstitial spaces are developed as courtyards. LEFT Exhibition rooms are punctured with low-level portholes, allowing natural light to filter in.

GROUND-FLOOR PLAN

Towada Art Center
Towada, Aomori Prefecture

OFFICE OF RYUE NISHIZAWA, 2008

The winner of an invited competition in 2005, Ryue Nishizawa's scheme for Towada Art Center aims to 'unify the experience of the city, the architecture and contemporary art'. The venue, which opened in April 2008, is a cluster of sixteen off-white cubes, loosely connected by curvy glazed walkways. Long vertical strips made of shiny steel sheets, which are 55 centimetres (21⅝ in.) wide but only 0.6 mm (1/50 in.) thick, uniformly cover the cubic structures. Their thinness and unevenness give an unexpectedly soft, fragile quality rather like that of rice paper, directly countering the robustness of the structure and austerity of the outlook.

As one part of the duo that makes up SANAA, Nishizawa has plenty of

OPPOSITE, TOP Paul Morrison's black-and-white mural *Ochrea* adorns the cuboid containing the cafe and shop. OPPOSITE, BOTTOM, AND BELOW The glazed walkway curves around such artworks as Noboru Tsubaki's *aTTa*, Jeonghwa Choi's *Flower Horse* and Yoko Ono's *Wish Tree for Towada*, which dot the site. RIGHT AND BELOW, RIGHT Also visible through the museum's apertures are Ana Laura Aláez's *Bridge of Light* and Ron Mueck's 4-metre-tall (13 ft) *Standing Woman*; at night, the museum is dramatically lit up by Kyota Takahashi's light artwork *Fragments of Color Cubes*.

experience in designing museums. Working independently, however, the architect has radically altered the way a contemporary museum operates. Here he snatches away the circular perimeter that neatly encloses a series of white cubical spaces in the 21st Century Museum of Contemporary Art in Kanazawa (pages 60–61), which he designed with Kazuyo Sejima. By scaling down the cubes to the street level as well as separating and spreading them out randomly on the site, which is on the town's spacious main avenue, he successfully liberates the new museum from any guardedness or aloofness, and connects it to the larger context: the city and its people.

Twenty-one notable international artists, including Ron Mueck, Yoko Ono, Do-Ho Suh, Michael Lin and Jim Lambie, were commissioned to produce works of art to be permanently installed at the centre. Some of the artworks spill out on to the front lawns and courtyards. During the day, the museum's exterior walls become the canvas for a large mural by Paul Morrison, and at night a screen drops for a light installation by Kyota Takahashi. The outside spaces are treated as part of the museum, and feel neither more open nor less closed than the spaces within.

Visitors are thus made to encounter works of art even before they enter the museum proper, and the passer-by inadvertently becomes the participant in Nishizawa's new 'village'. Large windows on the street side of some of the exhibition rooms and the double-height cafe and shop casually reveal the museum's activities. Conversely, they also frame the scenery as backdrops for the artworks exhibited inside. Even walking through its darkened rooms, visitors are constantly reminded of the close link between nature, the city and the art as they intermittently wander out into the open, sun-filled corridors, which whip up any stifling air that has collected inside the museum.

Yokosuka Museum of Art

Yokosuka, Kanagawa Prefecture

RIKEN YAMAMOTO & FIELD SHOP, 2002–2007

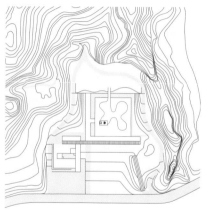

SITE PLAN

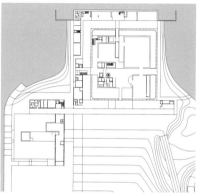

GROUND-FLOOR PLAN

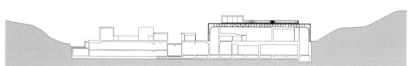

SECTION

Located at the entrance to Tokyo Bay, the city of Yokosuka has been a strategically important military port ever since Japan's capital was moved from Kyoto to Tokyo in 1600. The arrival in Yokosuka in 1853 of the fleet of American Black Ships headed by Matthew Perry effectively ended nearly two hundred years of isolation from the rest of the world, and marked the beginning of a new era for Japan. The city, which today has a US marine base, has therefore always been endowed with an outlook beyond itself.

Yokosuka Museum of Art faces the sea and is protected at the back by ancient forest, above which wheel magnificent black kites. Riken Yamamoto won the competition organized by the city according to a new system called QBS (quality-based selection), which examines the merit of each candidate's previous built works, rather than calling for anonymous entries. This system allowed Yamamoto, who had a good understanding of the area (his office is based in the neighbouring city of Yokohama) as well as a prolific profile of museum work, to shine, winning the hearts of the judges.

The trustees wanted the new museum to have a sufficiently strong appeal to pull in hip urbanites from the nearby metropolises, such as Yokohama and Tokyo, so the architect incorporated into the museum many features by which visitors could be entertained all day. The roof, for example, leads directly to a long, steep promenade through the forest. The main building is multi-layered: cuboid exhibition spaces are arranged in a curvy white steel structure punctured with large holes, enveloped within a rectilinear glass box, which acts as a shield from the harsh sea winds. The restaurant and museum shop are at the front of the building, sandwiched between the steel structure and the glass box, and diners get a full view of waves crashing on to the rocks and cargo ships passing by.

The spacious reception area is approached through one of the rounded openings, over a footbridge that crosses a deep-set cloister. The cloister scoops down into the basement, allowing natural light into the lower reaches of the building. The strategically positioned circular apertures also offer visitors perfectly framed glimpses of tranquility.

With plenty of glass throughout, the museum feels fluid and open; together with its annexe, it sprawls over 7000 square metres (75,350 sq. ft) of land and boasts more than 12,000 square metres (130,000 sq. ft) of total floor space.

ABOVE The curvaceous steel shell is encased in glass, for protection from the corrosive sea air. LEFT A view of the sea is framed through a porthole window. OPPOSITE, CLOCKWISE FROM TOP LEFT Natural light streams in through strategically placed round holes (LEFT AND CENTRE); a gallery; a large space in the basement reaches the full height of the museum; the entrance hall.

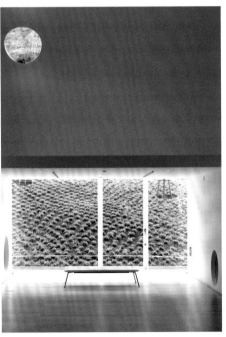

96	Brillare Dining and Party Room
100	Final Wooden House
102	Fujiya Inn
104	House of Light
106	Kuramure
108	Lamune Onsen
110	Miyagi Stadium
112	Moku Moku Yu
114	OPJ
116	Sapporo Dome
118	Spanish Pavilion/Aichi Expo 2005
122	Takasugi-an
124	Ware House
128	XXXX
130	Za-Koenji Public Theatre

Sport and Leisure

Brillare Dining and Party Room
Kobuchizawa, Yamanashi Prefecture

KLEIN DYTHAM ARCHITECTURE, 2005

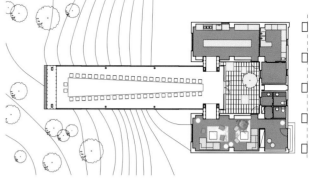

SITE PLAN

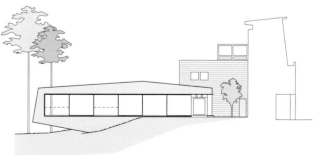

ELEVATION

Clad in shimmering stainless steel, Brillare is one of several recent, eye-catching additions to a resort complex with a brisk wedding business. Essentially it is a room for a table. However, this narrow tube jutting out into the forest is no ordinary room, and its 18-metre-long (59 ft) tapered table is hardly conventional either. But coupled with Leaf Chapel (pages 178–79), an earlier intervention by the same architects in the hotel's garden, it is the perfect place for a fairytale wedding.

Located near the entrance to the compound, Brillare is one of the first sights to greet visitors to the hotel. Partially cantilevered from the hillside, it extends into the woods at one end, and at the other plugs into an existing U-shaped building that originally housed one of several guest suites. The architects' first idea for the party pavilion was a Miesian glass box, but during the design process they transformed that pure, transparent volume into a solid one with large holes and an irregular profile.

By contrast with its striking exterior, Brillare's interior is defined by a clean, rectangular plan and section that echo Modernist architecture. A small courtyard abutting its narrow end

elevation leads the way in. Alternatively, guests may enter through one of the two tunnels that flank the room and anchor it to the lounge and an open kitchen, both in the U-shaped building. At each access point, the room is dramatically unveiled in one fell swoop.

Framed by floor-to-ceiling window walls running the length of the room, the expansive views are impressive, but even they cannot compete with the table. With space for twenty-two people on either side, the wedge-shaped slab gradually widens as it extends, directing the eye towards the focal point of the room: two special seats at the head of the table, where the bride and groom preside over their wedding feast.

Between the greenery outside and the guests in their wedding finery inside, there is no lack of visual richness. Except for the square-section columns, which are painted black, the finishes and furniture are bridal white. The only decorative element is a leafy motif that dances its way along the ceiling, sweeping down gracefully at the rear of the room to become a backdrop for the newlyweds.

In a country where weddings tend to be tightly choreographed affairs with little room for spontaneity, Brillare, the product of foreign architects with a fresh eye, breaks with formality and makes room for a little fun.

OPPOSITE, LEFT Brillare is a new, eye-catching addition to the existing Risonare hotel resort. OPPOSITE, RIGHT The shimmering stainless-steel room is cantilevered from the hillside. RIGHT The effect of the interior is light and sparkling. OVERLEAF An 18-metre-long (59 ft) tapered table is the room's defining feature.

SPORT AND LEISURE 97

Final Wooden House

Kumamura, Kumamoto Prefecture

SOU FUJIMOTO ARCHITECTS, 2005–2008

PLAN: LEVEL 0

PLAN: LEVEL 1

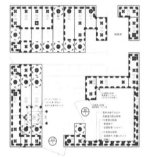
PLAN: LEVEL 2

PLAN: LEVEL 3

Final Wooden House is an experimental project won by Sou Fujimoto in 2005. The competition, which was run by Kumamoto Artpolis with Toyo Ito as the judge, required entrants to be under the age of thirty-five and the resulting project to be built entirely from timber (a material that could be freely provided by the client, Kumamura Forestry Association, which also operates a series of holiday bungalows in a valley in the lush countryside of Kumamoto on the island of Kyushu).

Although cedar is cultivated and commonly found all over Japan, the local forestry association still took some time to collect trees large enough to be cut into pieces with a 350-millimetre (13¾-in.) cross-section. These hefty pieces of timber have been piled up seemingly at random and discreetly bolted together to create a small bungalow about 4 cubic metres (140 cu. ft) in volume.

The dense presence of chunky timber, rough-cut and left untreated, gives the interior of the bungalow a sense of wildness. Users must negotiate their way past oversized steps and projecting beams as though they were still in the forest. In contrast, the flush exterior walls appear slick and neat, with a patchwork pattern created by the varying texture of the wood itself.

The distinction between walls, floors and stairs blurs, since any part of the structure can be used as a table, a chair or a bookshelf. Everything – apart from the small kitchen, shower and toilet, which are fitted under the steps to the left and right of the entrance – is created in wood. Users sleep cocooned among the rough but fragrant blocks: a unique and highly sensory experience.

With Final Wooden House, Fujimoto says he wanted to create a building

OPPOSITE, TOP The exterior view shows a sober wooden cube. Windows are cut in apparently random patterns. OPPOSITE, BOTTOM For the user, the building is more like a forest to be negotiated than a house. LEFT Any part of the wall or floor may be used as a table, a chair, a bookshelf or even a bed.

that was halfway between a cave and a house, to bring out the primitive within us. He likens such a creation to music without a score. A cube – the ultimate proof of Modernism's obsession with purity – has been dug out inside to reveal the toughness of the wood.

Toyo Ito correctly predicted that the structure would leak, but he also deemed that it did not matter. After all, people use it to be close to nature, and ultimately to think and move beyond the confines of normality.

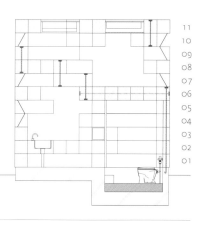

SECTION

Fujiya Inn
Ginzan Hot Spring, Yamagata Prefecture

KENGO KUMA & ASSOCIATES, 2005–2006

Ginzan Hot Spring, near the summit of a mountain in northern Japan, is a place where for centuries the Japanese have gone to rest and to reinvigorate themselves. The approach road gradually narrows and then divides into two paths either side of a small river crossed by many footbridges and flanked by *ryokan* (small, traditional, family-run inns); at its end is a waterfall.

Kengo Kuma had previously worked there, not far from the inn, moulding a public bathhouse from dark native timber. Although modern, its understated palette is effortlessly assimilated into the scenery. Impressed by what he had done, Jeanie Fuji – Japan's only American *okami* (keeper of a traditional inn) – and her husband, Atsushi, whose family has run a *ryokan* here for several generations, commissioned Kuma to modernize their 350-year-old inn.

Despite the strict scenery-protection policy enforced in the area, the architect dispensed with much of the past. All that is left of the old *ryokan* are its traditional roof line and some of its original beams. Although such traditional materials as paper, bamboo, stone and timber are plentifully used to help assimilate the past, they all in fact work to undermine, rather than reinforce, the criteria of the preservationists.

Kuma, who is one of Japan's most vocal architects (he has written many books and regularly contributes to newspapers and magazines), makes broad stylistic references. The wooden latticework that characterizes traditional *machiya* (merchants' houses), for example, covers much of Fujiya's façade in amplified form, echoing other buildings in which Kuma uses louvres to soften façades of glass or steel. The hand-blown stained glass, made specially in France, sets off the threshold as if it were another river that visitors must cross.

Verticality is emphasized throughout the three-storey timber-framed building. Some 1,200,000 bamboo strips, each hand-cut to just 4 millimetres (⅛ in.) wide and strung up with tiny pins by master carpenter Hideo Nakata and his son, from Kanazawa in Ishikawa Prefecture, liberate the reception area from the heaviness of its grey granite floor. Collaborating with other craftspeople, Kuma succeeds in bringing fluidity, warmth and excitement to the countryside.

ABOVE Fujiya Inn follows the contour of the old inn to preserve the so-called *Taisho Roman*, the 'romance' of the Taisho era (1912–26). OPPOSITE, LEFT, TOP AND BOTTOM Bedrooms are kept minimal. OPPOSITE, TOP RIGHT Subtle lighting and floating stairs are Kuma's forte. OPPOSITE, BOTTOM CENTRE Changing area for one of the spas. Verticality is emphasized by strips of bamboo. OPPOSITE, BOTTOM RIGHT One of five hot springs at the inn.

SITE PLAN

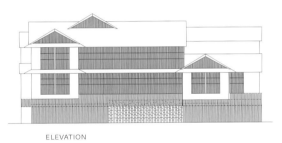
ELEVATION

GROUND-FLOOR PLAN

House of Light

Tokamachi, Niigata Prefecture

JAMES TURRELL WITH DAIGO ISHII + FUTURE-SCAPE ARCHITECTS, 2005

House of Light is a collaboration between a Japanese architectural team and the US artist James Turrell. The two-storey guest house perched high on a hillside was designed for two groups of families or friends to stay in comfortably, and encourages them (with its retractable roof) to appreciate the intensity and subtle colour changes of the sky. The roof typically slides open an hour before sunrise and sunset, and the building's rural location and its position high above the town of Tokamachi mean that nature's splendour can be appreciated with little acoustic disturbance. Indeed, the building was conceived more as a meditation centre than as either an artwork or a hostel.

Suitably, then, the building's deep eaves and strong processional central stairway, which provides access direct to the upper level, recall the temple vernacular. A broad cloister circumnavigates the upper storey. The house is simply and elegantly divided into two wings: one

containing the first-floor kitchen and eating facilities and a ground-floor *tatami* room; the other a room with the retractable roof over a spacious hot spring. The way the thresholds and proportions are determined by the *tatami* mats is highly traditional and re-creates the serenity of bygone eras.

Turrell works in a variety of ways, using both natural and artificial light, but one of his defining characteristics is an ability to make light appear solid: light-filled voids can often be mistaken for solid walls, and rooftop openings are crafted not to provide a source of light but to encourage visitors to gaze at a panel of sky, which can sometimes seem to have a more material presence than the room itself. Careful backlighting around the upper edges of the rooms also provides a neutral backdrop against which subtle celestial colour changes can be discerned.

The artist and architects together determined light settings throughout the House of Light. Most remarkable is the lighting in the hot spring, for there is none, apart from strips of ultraviolet light that mark out the edges of the room. Since one must bathe in almost complete darkness when night falls, focusing on the sensual pleasures the hot spring gives, the architecture disappears entirely.

OPPOSITE, TOP The main approach to the guest house. OPPOSITE, BOTTOM The retractable roof allows the creation of various lighting schemes. ABOVE The sky, seen through the skylight, appears as a slab of pure colour. LEFT In the lower-level spaces, artificial light is also treated to gradations of intensity.

Kuramure

Otaru, Hokkaido

MAKOTO NAKAYAMA, 2001–2002

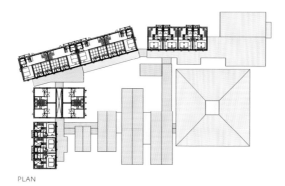

PLAN

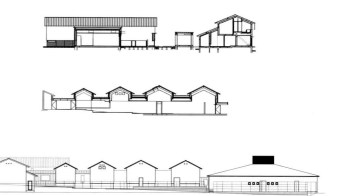

SECTIONS AND ELEVATION

Built on the edge of the city of Otaru – famous for its turn-of-the-century warehouses and the gaslights in its old port – in a landscape of hills and hot springs, this *ryokan*, a uniquely Japanese hotel with a particular emphasis on privacy, reinterprets this popular tourist destination in a contemporary way. Its name reflects its arrangement: huts resembling old storehouses (*kura*) in a cluster (*mure*) are linked by discreet passageways and halls. This luxury resort of nineteen suites (eleven maisonettes and eight two-storey flats) – with facilities including private dining rooms, hot springs, a bar and a library, as well as a small gallery that turns into a wedding chapel – wraps around a courtyard while providing particularly well-crafted views of the landscape beyond.

The complex consists of simple pitched-roof buildings of concrete and steel with traditional motifs, evocative of the warehouse or agricultural vernacular. These structures are often sliced off at the bottom, providing low-level strip windows that align with the ground plane. Small windows are also cut to provide glimpses of particular views or niches for objects. Moreover, the textured exteriors of modern industrial materials contrast with the lofty ceilings, stone floors and oiled timber of the interiors.

Respect for privacy lies at the heart of the architectural programme here. The entrances to the guest rooms are recessed, affording visitors the luxury of taking off their shoes in private rather than at the lobby, as is the case in most *ryokan*. Dining rooms are also enclosed, and each has a recessed entrance, much like the guest rooms. The subtly undulating low-lit corridors pose as the public domain, although they also have a protective layer of privacy. Strollers

never have to acknowledge the busy road outside or other hotels near by.

All views are strictly controlled to create a microcosm of pebbles, rocks and plants, of which even the distant landscape beyond becomes a part. As in Kengo Kuma's Fujiya Inn in Ginzan (pages 102–103), the architect's particular feel for materials and textures becomes important in creating the theatrical illusion. Some walls, for example, are moulded in lead, creating bubbly, uneven, rustic surfaces. The discreet signage and the momentary lapse into darkness into which visitors are thrown after the heavy front door gently slides open signal a way into this dense and shadowy space, rich in patina.

OPPOSITE, TOP LEFT AND RIGHT Gabion walls add texture to the development. OPPOSITE, BOTTOM Low-level strip windows align with the ground plane. ABOVE, LEFT Personal dining rooms are enclosed, with recessed entrances. ABOVE, RIGHT A retractable wall unifies spa and landscape. LEFT The interior deploys an architectural language of intersecting and overlapping planes. Here, visual depth is provided by a view into the courtyard.

Lamune Onsen

Takeda, Oita Prefecture

TERUNOBU FUJIMORI, 2004–2005

LEFT, TOP Exterior view. The walls are clad in bands of charred cedar wood and white mortar. LEFT, BOTTOM Entrances to individual family bathrooms. OPPOSITE, CLOCKWISE FROM TOP LEFT Live pines spring from the steam towers; this *rotenburo* (outdoor bath) reserved for women is filled with the region's unique warm, carbonated spring water; a wall in the men's bathroom is embedded with mother-of-pearl shells; Lamune Onsen is open to the public, as this sign shows.

With Lamune Onsen ('Lamune hot spring house'), Terunobu Fujimori experiments with the language of the tower. Specially constructed towers allow steam to escape, but, not satisfied with simply building functional structures, Fujimori added living pine trees – associated in Asia with vitality and long life – which spring out of them.

Although he originally trained as an architect, Fujimori has taught architectural history at Tokyo University for most of his career, and only recently began to practise as an architect. Perhaps this new discipline has permitted him to explore the less serious and more youthful side of his nature; certainly, his projects are characterized by eccentricity and humour.

Lamune Onsen stands beside a small river in the idyllic town of Takeda in southern Japan, a short walk from the owner's traditional *ryokan* (inn), which has been in his family for generations. The name of the bathhouse, which is open to the public and not only to guests at the *ryokan*, is derived from a fizzy drink popular with children in Japan. It resembles the warm, carbonated spring water that is unique to this region, and which is available for guests to bathe in, in addition to the hot spring water that is found all over Japan.

The bathhouse, clothed in alternate strips of charred cedar and white mortar, with its towers capped in hand-rolled copper sheets and pine trees, has a fairytale appeal. The waiting-room chairs are carved from charred cedar, and one wall in the men's bathroom is embedded with mother-of-pearl shells. In a playful touch, a small window flanking the gallery is shaped like a mammoth.

The writer Junichiro Tanizaki, who lamented the loss of traditional Japanese aesthetic sensibilities in his book *In Praise of Shadows* (1933), would have been pleased by the patina of the mortar walls, weathered by the steam. It is as though the bathhouse has been there all along, untouched for centuries but suddenly animated by the kiss of a Prince Charming.

SITE PLAN

ELEVATION

Miyagi Stadium
Rifu, Miyagi Prefecture

ATELIER HITOSHI ABE, 1994–2000

For centuries, the geometry of concentric circles has dominated stadium construction. But architect Hitoshi Abe knew there had to be a better and more open venue for sports lovers to cheer on their favourite teams. Completed for the Japan/Korea World Cup of 2002, Miyagi Stadium proves his point.

A dramatic win in an international competition – and a commission that got his career off to a flying start – Abe's 49,000-seat arena is the star of a large athletics complex on the outskirts of Sendai, the architect's home town. Taking its cue from the site's hilly topography, the stadium consists of two open-ended, asymmetrical grandstands arched around a playing field. While the dynamic, double-tiered main stand soars above the sloping site, the single-tiered back stand sinks into it, fusing architecture and landscape.

Accessible from all directions, the stadium has various entrances, but most spectators ascend the massive concrete ramp connecting the main entrance to the surrounding circulation concourse, a wide corridor that rings the third floor of the main stand and the top of the back stand as well as the exposed seating areas between. From here fans ascend or descend to their seats. Administrative offices and assorted training facilities are underneath the stands.

Complementary but independent crescent-shaped roofs cover the two grandstands. While the rear roof hovers protectively above the seats, the main roof arches up gracefully and touches down beyond the building, where it terminates in a massive concrete block at either end – the stadium's most impressive feature. The blocks appear to secure the swooping cover without breaking sweat, but such a feat requires an enormous keel truss to hold up the main roof, and a buried reinforced-concrete beam to pull the ends of the truss together like a crossbow. A line of concrete abutments, no two the same, props up the roof from below.

In addition to the formal connection between Miyagi Stadium and its site, Abe envisioned blurring the functional boundaries between park and arena. While the underbelly of his building includes athletics facilities (many of them designated for public use), a running track, also intended for local citizens, races around the stands. This creative use of space that is often overlooked or underused gave Abe the edge over his competitors.

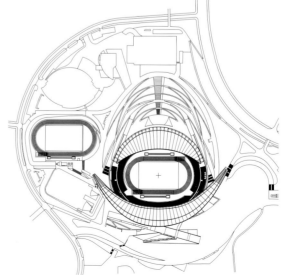

SITE PLAN

ABOVE The slick new stadium in the outskirts of Sendai was built for the 2002 World Cup. OPPOSITE, TOP LEFT A line of concrete abutments, all different, supports the roof of the main stand. OPPOSITE, TOP RIGHT Stairs lead up to the main pedestrian concourse. OPPOSITE, BOTTOM LEFT The roof over the main stand terminates at both ends in massive concrete blocks. OPPOSITE, BOTTOM RIGHT All elements, including the seats, are aligned perfectly.

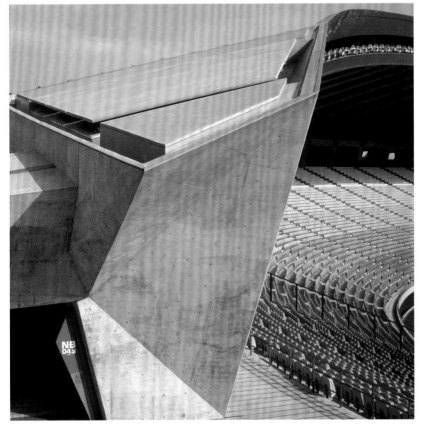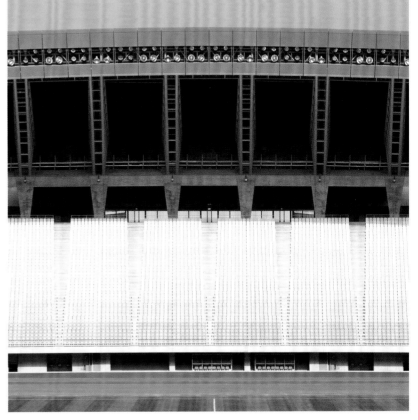

SPORT AND LEISURE 111

Moku Moku Yu

Kobuchizawa, Yamanashi Prefecture

KLEIN DYTHAM ARCHITECTURE, 2004–2006

A communal bathhouse shaped like a collection of water droplets, Moku Moku Yu sits in a clearing in the grounds of a hotel complex designed by Mario Bellini during Japan's economic bubble period. The new amenity is one of several interventions by Klein Dytham Architecture intended to spruce up the resort. Although the firm's Brillare Dining and Party Room and Leaf Chapel (pages 96–97 and 178–79) turned the hotel into a popular wedding venue, the complex lacked that standard feature of Japanese destination resorts: the communal bath. Moku Moku Yu has solved that problem.

A popular mode of relaxation in Japan today, the bathing ritual is steeped in history. While the architects were determined to make a clean break from tradition, their first image for the project was the barrel-shaped wooden tub in which Japanese people have soaked for centuries. Expanding on that idea, the architects created a cluster of seven intersecting circles, each one for a different stage of the bathing process.

Accessed from the main building by a footpath, the bathhouse complex begins with a reception area, a free-standing round room where guests pick up towels. From there men and women proceed to separate cylindrical changing rooms, lined with banks of lockers and sink counters that radiate from the centre of the room. Like Venn diagrams, the changing rooms overlap with a second pair of circular rooms, still divided by gender, where bathers scrub down before entering the tub. Supporting a row of wall-mounted showers, a sweeping, curved wall cuts across the room, partitioning the bath. Brimming with steamy water heated to muscle-massaging perfection, the two tile-lined pools, one for men and the other for women, are purely for relaxation. Offsetting the bath's intense heat, cool mountain breezes waft through the building's semi-permeable skin of solid wooden staves stained pink and blue, interspersed with unglazed openings.

Those who crave more exposure to the elements may move on to the shared outdoor pool. Uniting the male and female facilities, it is one of the first mixed baths to be built since the practice (outlawed in the 1940s) was re-legalized. Bathers must still drape themselves with towels, however. A fence of tree trunks offers privacy and blurs the boundary between the man-made and the natural.

The bathhouse's name is an apt play on words (*moku* means 'wood', but *moku moku* means 'plumes of steam'). More than just a catchy phrase, the name embodies the ingenious spirit of Klein Dytham's architecture.

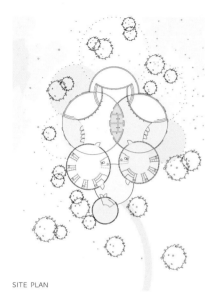

SITE PLAN

OPPOSITE, TOP Connected to the hotel by a footpath, the communal hot springs are shaped like a collection of water drops. OPPOSITE, BOTTOM The *rotenburo* (outdoor bath) follows the circular contour of the complex. BELOW Stained pink and blue, colourful wooden staves form the exterior of the buildings.

SPORT AND LEISURE 113

OPJ

Irabu, Okinawa Prefecture

JUN AOKI & ASSOCIATES, 2006–

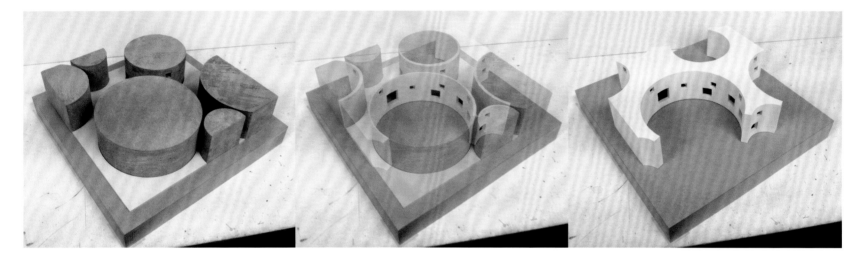

Irabu is an even smaller island than Okinawa, off which it lies. Tropical and removed from civilization, it was the ideal place for a luxury resort of the type that once attracted Japanese tourists in droves. With that goal in mind, a developer invited Jun Aoki to try his hand at hotel design.

On visiting the island, Aoki was deeply impressed with the rugged beauty of its rocky landscape. As with all the islands in the area, Irabu is composed largely of Ryukyu limestone formed from coral from the surrounding sea. An inherently weak material vulnerable to the ravages of rain and seawater, the porous stone is riddled with circular craters of all sizes.

Inspired by the dramatic site, Aoki first intended to incorporate the geology into his architecture by turning its spherical formations into habitable spaces. The client gave the idea a thumbs up, since the property sits within a nature reserve and the scheme was subject to intense scrutiny by local authorities. But as the idea evolved, Aoki decided to invert solid and void by designating the circular depressions as private outdoor spaces and using the interstitial areas for indoor functions.

Blending with the scenery, the resort consists of an entrance hall, a restaurant, a library and fifteen individual guest suites. The complex's unmonolithic, free-form shape follows the coastline, with man-made pieces partially embedded in the ground.

Although the guest suites are all configured differently, Aoki designed them with the same criteria in mind. Each consists of 100 square metres (1075 sq. ft) of interior space and 400 square metres (4300 sq. ft) of exterior space in three circles. Between the ocean and the sleeping quarters, the largest circle holds a private pool and hut. The bedroom also abuts the medium-sized circle, which contains a tropical garden at the entrance. The smallest circle holds the jacuzzi and connects to the indoor bath.

To create a cave-like ambience, Aoki kept refinement and decoration to a minimum. To complement the floor's irregular surface, he specified *hatsuri*, a finishing technique of roughly hammering masonry surfaces and coating them evenly with white paint, for the walls. He was after a natural look, but achieving such a goal would require precise calculations and imported craftsmen from Kyushu.

Unfortunately, this ambitious project was an early casualty of the economic downturn in 2008. But Aoki's innovative concept is bound to resurface at a later date.

ABOVE Study models. OPPOSITE, CLOCKWISE FROM TOP LEFT Circular depressions in the ground hold outdoor functions, while the interstitial spaces hold indoor functions; conceptual model; conceptual drawing; site rendering.

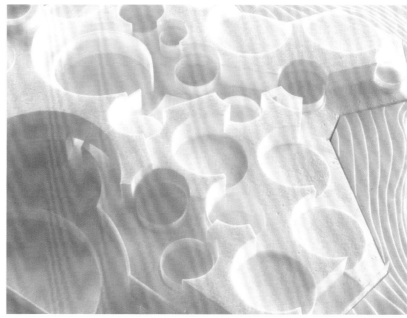

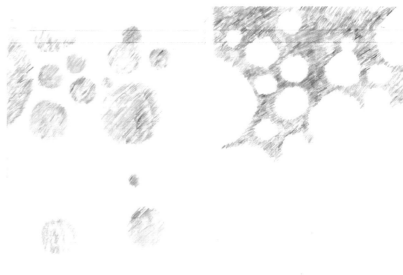

SPORT AND LEISURE 115

Sapporo Dome

Sapporo, Hokkaido

HIROSHI HARA, ATELIER φ AND ATELIER BNK, 1998–2001

While Sapporo's snowy climate was a boon for the 1972 Winter Olympics, it was less suitable for the provision of the regulation turf for World Cup soccer. When the city was chosen as a host for the 2002 tournament, the municipal government kick-started the event with a design competition for a new stadium to seat 43,000 people. Toppling entries from around the globe, Tokyo architect Hiroshi Hara's team (including construction companies Taisei Corporation and Takenaka Corporation) was the winner.

As well as the football association's requisite grass field, the city wanted the stadium to have a professional baseball diamond and to be able to accommodate rock concerts and exhibitions. Neither the programme nor the site – flat farmland on the outskirts of town – posed any significant problems for the designers. The real kicker was the climate: during the long winters in Sapporo, the mercury seldom rises above freezing, Siberian winds blast mercilessly and the snowfall is enormous.

The challenge facing the designers was to protect the grass yet provide sunlight to keep it healthy. The obvious solution – a retractable roof – was trounced by Sapporo's heavy snows. The only option was a moveable playing field.

In principle, the solution is very simple. Shaped like a figure of eight, the stadium's two arenas (one exterior, one interior) share a playing field that slides between them. While the outdoor half is essentially a depression in the earth, the indoor half is covered with a shimmering stainless-steel shell resembling a flying saucer. Creating column-free space inside, a 4-metre (13-ft) truss supports the dome and transfers its weight to perimeter columns, except at the open end, where a steel beam and tension cables tie the system together without blocking passage.

Although Hara's solution seems otherworldly, it is rooted in tradition. The dome is equipped with a kit of adjustable parts, like the temporary furnishings used to adapt historic houses to the changing seasons. Synthetic groundcover and a false pitcher's mound convert it into a baseball stadium. Most of the seating

folds and moves to turn the stadium into an event hall. And the entire ground plane – steel, concrete, sand and grass – glides magically, enabling the soccer field to stay green. The five-hour conversion process begins with reducing the weight of the field by blowing and trapping air underneath with perimeter seals. The field then moves 4 metres (13 ft) a second with the aid of pneumatics. Once inside, the field is rotated 90 degrees to face the main stand, the seats are locked into place and the green is again ready for play.

TOP The stadium hovers amid flat farmland on the outskirts of Sapporo. ABOVE A steel truss supports the roof, yielding column-free space. OPPOSITE, TOP LEFT Covered with gleaming stainless steel, the dome resembles a flying saucer. OPPOSITE, CENTRE LEFT The playing field slides in and out of the building through the glazed end wall. OPPOSITE, TOP RIGHT AND BOTTOM Roof details. The cylindrical viewing bridge pierces the dome, offering an aerial view of Sapporo as well as of the stadium's interior.

SPORT AND LEISURE 117

Spanish Pavilion/Aichi Expo 2005

Nagakute, Aichi Prefecture

FOREIGN OFFICE ARCHITECTS, 2004–2005

With its emphasis on sustainability, Aichi Expo 2005 was a decidedly abstemious affair in contrast to Asia's first international exposition, in Osaka in 1970. To make as little environmental impact as possible on the parkland of the site, guest countries were assigned prefabricated 18-square-metre (194-square-foot) sheds, which could be dismantled easily once the show was over. Unfortunately, many countries failed to engage fully and imaginatively with the topic of sustainable development, using the sheds merely as backdrops for projections advertising their best beaches, lakes and mountains, or as corner shops selling souvenirs and food.

The Spanish Pavilion was a rare exception to the general passivity. Designed by Foreign Office Architects (FOA) – which was co-founded by a Spaniard, Alejandro Zaera-Polo – its structure encapsulated that country's dynamic cultural identity in the modern, international world. The architects' view was that Spain's unique heritage has sprung from a fusion of Judaeo-Christian and Islamic cultures. FOA began to play with a wide range of architectural elements, such as courtyards, screens, cloisters, geometric patterns, lattices, tracery and arches, in order to generate a form that might represent the country without becoming a mere cliché.

The cladding solution not only conveyed the idea of all things Spanish but also acknowledged the local craft for which this region of Japan, particularly Seto, is famous: the production of ceramics. Spain, of course, has its own expertise in ceramics, and taking its cue from the patterns present in Moorish design, FOA designed an architectural skin comprising six types of distorted hexagon, each in a different shade of red or yellow to represent anything from the country's national flag to bullfights, sun, sand and wine. Bolted on to a steel sub-structure, each 6-kilogram (13-lb) tile was given a specific place and orientation on the façade, and the pattern never repeats itself, producing a continuously varying arrangement of geometry and colour. The use of ceramics, moreover, symbolized the physical carrying of Spanish earth to Japan.

The pavilion came to be configured as a central space around which seven smaller exhibition zones were placed, in much the same way as small chapels are accessed from the nave of a large church. The order in which visitors would enter each space was not predetermined, other than that the

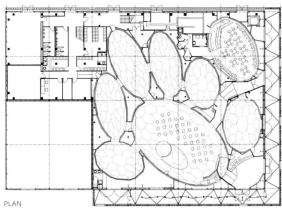

PLAN

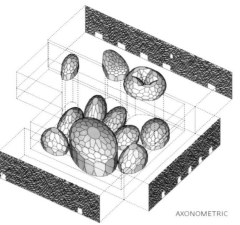

AXONOMETRIC

OPPOSITE, TOP The Spanish Pavilion was clad in ceramic forms suggestive of Gothic latticework and Islamic patterns. OPPOSITE, BOTTOM Ceramic hexagons were bolted to a steel sub-frame. BELOW Within the pavilion, a central space (BOTTOM RIGHT) provided access to smaller, dedicated exhibition zones, in much the same way as chapels are accessed from a nave.

SPORT AND LEISURE 119

Kaisho Forest Observation Tower is one of few survivors from the Expo. Assembled in the traditional manner, it uses no mechanical fixtures.

120 NEW ARCHITECTURE IN JAPAN

central, enclosed 'courtyard' would be the vehicle by which the themed 'chapels', each structured as a vaulted bubble, would be visited. 'Ornate gothic vaults, Islamic domes and faceted vaults are reinterpreted as more free-form structures', the architects explained.

Little survives now from the Expo: only the Ferris wheel, a museum and Atsushi Kitagawara Architects' Kaisho Forest Observation Tower, the tall latticework of which was constructed using a traditional method of wood joinery that requires no mechanical fixing. The site reopened in July 2006 as Aichi Expo Memorial Park (known as Morikoro Park), containing an ice rink and other municipal facilities. Work on the park is ongoing, however, including the construction of a large exhibition hall and sports centre, the Global Citizen Communication Centre, designed by Atelier Bow-Wow. Scheduled for completion in 2010, this 8850-square-metre (95,300-sq.-ft) building promises a long, sweeping, elevated approach route along its grass rooftop to meet a dramatic geodesic structure containing a gymnasium.

TOP LEFT Site plan of the new Morikoro Park. TOP RIGHT Atelier Bow-Wow's vision for the park is seen from above. ABOVE The internal spaces of the proposed public facilities are large, and include a geodesic dome over the sports hall.

SPORT AND LEISURE 121

Takasugi-an

Chino, Nagano Prefecture

TERUNOBU FUJIMORI, 2003–2004

BELOW, LEFT In keeping with tradition, the hearth is placed in a corner. RIGHT, TOP A large window frames a view of Fujimori's home town. RIGHT, CENTRE Fujimori prepares tea in his tea house. RIGHT, BOTTOM, FROM LEFT The gilded light well seen from below; one of the load-bearing tree branches elbows in. OPPOSITE Takasugi-an hovers high above the ground.

The academic and architect Terunobu Fujimori has observed that a tea house is 'the ultimate personal architecture'. Its extreme compactness, which would at most accommodate four and a half tatami mats (2.7 square metres/ 29 sq. ft of floor space), makes it feel 'more like a piece of clothing'.

Following the tradition where tea masters maintained total control over the construction of these simple enclosures, Fujimori decided to build a humble tea house for himself and by himself on a patch of land owned by his family. His interest as an architect, however, lay more in pushing the limits and constraints of a traditional tea house than in pursuing the art of tea-making itself, and, as a result, he has created a highly expressive piece of architecture.

Takasugi-an, which literally means 'a tea house [built] too high', is indeed more like a tree house than a tea house. Guests enter by climbing the ladders propped against one of the two chestnut tree trunks supporting the structure. The trees were cut from a nearby mountain and brought to the site. Guests take off their shoes on a platform just over halfway up. Inside the room, with its simple finishes of plaster and bamboo mats, the architect's adventurous spirit gives way to the serenity more suited to the purpose of making tea and calming one's mind.

A large window frames a perfect bird's-eye view of the town where Fujimori grew up. It effectively replaces *kakejiku* (picture scrolls), which in the traditional tea house would give clues to the time of year. Alas, this *kakejiku* displays not only the cyclical seasonal changes but also the profound and irreversible changes taking place in such provincial towns as Chino. Also visible from the large window is Fujimori's very first project, Jinchokan Moriya Historical Museum.

What looks like the chimney in this tea house is in fact a periscope-like light well, gilded to reflect and amplify the gentle light of the setting sun. In keeping with tradition, the hearth is in one corner with the chimney above it. The roof is covered with what is fast becoming Fujimori's signature material: hand-rolled copper sheets. The architect's penchant for the personal, vernacular and everyday is particularly evident in this tea house, which sways gently like a mother holding a newborn baby in her arms.

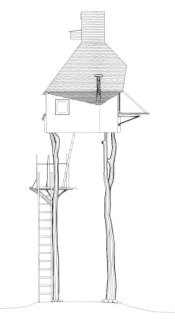
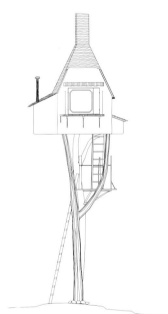

ELEVATIONS

FLOOR PLAN

SPORT AND LEISURE 123

Ware House
Asahikawa, Hokkaido

JUN IGARASHI ARCHITECTS, 2006–2008

Located in the wilds of Hokkaido, where winters are fierce and snowfall abundant, Ware House is essentially a weatherproof receptacle for four cars. A garage and gallery combined, it was commissioned by a client with a taste for fancy wheels and a wish to bring his collection together in one climate-controlled concrete box.

Despite this unusual programme, the Hokkaido-based architect Jun Igarashi did not have to look far for inspiration. He took his design cues from the century-old stone storehouse that shares the site with the client's more recent, timber-framed house. The client – a consummate collector who works in the recycled-goods business – purchased the antique structure and resurrected it on his land. While the old building handily accommodates his eclectic assortment of antiques and collectibles, his vintage Cadillac, streamlined Lamborghini and muscular BMW needed a building of their own.

Modelled on the historic storehouse but four times bigger, Ware House is also made of fireproof masonry walls – in this case, extra-thick concrete. In a further quotation from tradition, a grid of beefy wooden beams, each a giant 1-metre-deep (3 ft) slab of local pine, supports the roof. This configuration yields a cavernous, column-free space where the cars are artfully displayed.

A large sliding door at ground level provides an entrance for the cars, but visitors descend a short run of stairs before entering the building. Wrapping the central void are three narrow galleries, one below grade and two above, enabling the owner to admire his cars from various vantage points. Ladder-like steel stairs connect the levels without impeding sightlines, but – should the client need a break from his prized possessions – there are also a kitchen, a shower and a Japanese-style, *tatami*-floored room for snoozing or sipping tea.

Because the client wanted the natural light levels kept down, Igarashi limited windows to strips of glass at floor level and a clerestory band. The resulting shadowy interior, coupled with the stark grey surfaces, brings parking lots and storage vaults to mind. But against the backdrop of raw concrete, the Caddy's chrome detailing and the Lamborghini's sleek silhouette are even more stylish.

LEFT Made of extra-thick concrete, Ware House is built to last at least as long as the cars. OPPOSITE, TOP LEFT A strip window at floor level increases the drama and restricts the amount of incoming sunlight. OPPOSITE, TOP RIGHT Unobtrusive staircases keep the space free of visual clutter. OPPOSITE, BOTTOM LEFT The cars gleam against the raw concrete of the walls. OPPOSITE, BOTTOM RIGHT Le Corbusier Cube chairs complement the black of the BMW and the Cadillac. OVERLEAF The wooden grid supporting the roof softens the glare from the sun, creating an atmospheric enclosure for the owner to view and enjoy his treasured collection.

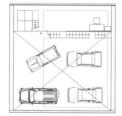

FLOOR PLANS

SPORT AND LEISURE 125

XXXX

Yaizu, Shizuoka Prefecture

MOUNT FUJI ARCHITECTS STUDIO, 2003

RIGHT The new ceramics studio was built by the architects and their friends. Sheets of laminated plywood form the floor, walls and ceiling. OPPOSITE Enclosed with window walls at either end, the studio faces a public park in one direction and the client's garden in the other.

For $15,000 one can purchase a new car complete with electric windows, air conditioning and built-in navigation system. Such a sum does not go far when it comes to construction, yet when a shipbuilder-turned-ceramicist decided to blow his savings on a one-room studio in his back garden instead of a car, that was the budget. Unfazed by the extreme financial constraint, Mount Fuji Architects Studio was quick to take on the challenge.

To make the job run smoothly, however, the architects needed a streamlined approach to design and, especially, construction. Because the funds would not even cover the cost of a general contractor, the architects' only option was to build the studio themselves with the help of able-bodied friends willing to sacrifice a few days of holiday. But this approach meant they had to devise a simple assembly method that did not need much input from experts. Nor could it require heavy machinery that the budget would not support and which the architects were not licensed to operate.

The most economical solution was to build with sheets of standard-issue plywood that could act as structure, skin, insulation and interior finish simultaneously. By laminating four or five pieces with glue, the architects were able to convert the wood planes into wall, roof and floor panels. Traditional joinery methods turned the two-dimensional planes into eight three-dimensional but non-rectangular frames. And pinning the frames together at their cross-points – half of them lean to the right and half to the left – resulted in an efficient as well as elegant 22-square-metre (237-sq.-ft) tube of space with a unique cross-shaped profile.

Once these basics were in place, the architects called on a glasscutter and a door carpenter to add the finishing touches. While full-height panes of glass were needed to close off the ends of the studio, smaller pieces were tailored to fill the triangular gaps where the frames do not align. Conventional metal clips and carved grooves in the wood are all it took to hold the transparent sheets in place. The two doors were another matter: set into the inclined panels on the south side of the building, both had to be hinged top and bottom to swing open. Finally, the architects coated the exterior with nautical-grade waterproofing to keep the wood looking and behaving its best.

Full of daylight and facing greenery – one end looks on to the client's garden and the other faces a public park – the no-frills handmade studio suits a potter who throws his own cups as well as more ambitious pieces for sale. If, in the future, he chooses another addition over an Audi, all he has to do is attach a further 'X' or two.

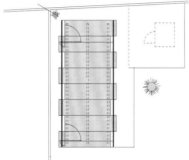

SITE PLAN

SECTIONS

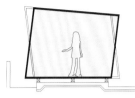

CONCEPTUAL DIAGRAMS

SPORT AND LEISURE 129

Za-Koenji Public Theatre

Suginami, Tokyo

TOYO ITO & ASSOCIATES, ARCHITECTS, 2006–2008

BELOW Private houses and small apartment buildings surround the theatre. OPPOSITE, TOP Dotted with circular lights, the internal stair is a place to see and be seen. OPPOSITE, BOTTOM LEFT The lobby. OPPOSITE, BOTTOM RIGHT The cafe tops off the building.

Capped with a billowing roof that rises and falls dramatically, Za-Koenji evokes the excitement of a big top. Its dynamic, attention-grabbing form, clad entirely with chocolate-coloured steel plates and sprinkled with circular windows, beckons spectators invitingly. It is the product of Japan's architectural superstar, Toyo Ito, who won the commission in an invited competition. Modest in scale and targeted to local audiences, this fantastic building may not house 'the greatest show on earth', but it is by far the leading attraction of Tokyo's Koenji neighbourhood.

On the fringes of the city and known for both its vibrant youth culture and its annual Awaodori traditional dance festival, Koenji is far removed from the city's main cultural attractions. Instead, its cluttered streets are lined with dry-cleaners' shops, convenience stores, car parks and the other trappings of daily life. Just a stone's throw from the closest station hub and facing an elevated railway line with commuter-laden carriages whipping past every couple of minutes, Za-Koenji nevertheless greets the street warmly with an open plaza.

Inside, the mundane surroundings are immediately left behind and the magical world of the theatre takes over. The process begins in the main lobby, which is animated by reflected discs of light that dance across the floor. The six-storey building has three performance spaces stacked on top of one another. Za-Koenji 1, a 230-seat specialized theatre for trained professionals, is on the ground floor, while Za-Koenji 2, on the second basement level, is a 300-seat auditorium for general use. It abuts Awaodori Hall, a multi-purpose space intended for informal presentations and rehearsal of the dance that has made Koenji famous. Support spaces – dressing rooms, costume and prop studios, and a script archive – surround the theatres, and a graceful staircase connects the levels, culminating with dramatic flourish in the cafe, topped by the wavy roof.

A careful choreography of internal height requirements and external height restrictions, the roof's remarkable profile is a collage of curved sections taken from seven different conical or cylindrical forms. Welded seamlessly, they swell upwards towards the railway lines but fall back down at the rear, as if bowing politely to the primary school and the private houses at the back of the site. They also blend with the exterior walls, resulting in a single enclosure. Some 230 small circular apertures are positioned according to interior function, filled with flush-mounted, milky glass panes that twinkle like stars and capture the imagination.

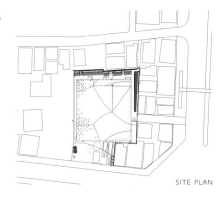

SITE PLAN

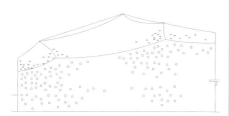
ELEVATION

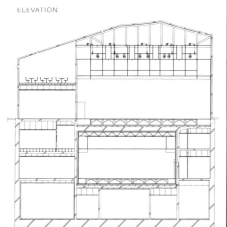
SECTION

SPORT AND LEISURE

Education

134 Bubbletecture M
136 Fuji Kindergarten
138 Fukusaki Hanging Garden
140 Fukutake Hall, Tokyo University, Hongo Campus
142 Future University – Hakodate
144 Gifu Academy of Forest Science and Culture
146 Gunma Kokusai Academy
148 Ina-Higashi Elementary School
150 Kaito Workshop
152 Minami-Yamashiro Primary School
154 Mode Gakuen Cocoon Tower
158 Mode Gakuen Spiral Towers
160 Musashino Art University Library
162 Rikuryo Alumni Hall, Kitano High School
164 Tama Art University Library

Bubbletecture M

Maihara, Shiga Prefecture

ENDO SHUHEI ARCHITECT INSTITUTE, 2001–2003

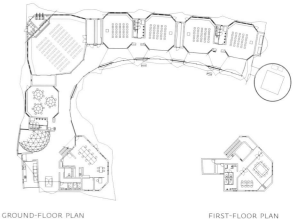

GROUND-FLOOR PLAN   FIRST-FLOOR PLAN

ELEVATIONS

This municipal nursery school propelled its then little-known architect on to the international architectural stage. The undulating, larger-than-life building is constructed from timber, steel and reinforced concrete, and contains four classrooms, a large playroom, a library and plentiful 'intermediary' space in which children can run around. Its location, in a new town, meant that the architect had little local context to which to respond, so he was free to create a form that acts as its own landmark.

The structure contains strong references to the geodesic domes developed by the US designer and thinker Buckminster Fuller (1895–1983), who demonstrated that robust spherical and hemispherical forms can be assembled from flat-sided triangles. The kindergarten is not strictly a geodesic structure, since it flows gently across the site and requires propping up at regular intervals, but it does contain a genuine geodesic dome, which houses the library: a dome within a dome.

Shuhei Endo is known for his high-tech approach. Here, since timber was specified by the client, he assembled pieces of wood with hexagonal metal fixtures that could be adapted into 284 variations. Computer-aided design and production scheduling allowed the components and assembly to be easily tracked and coordinated. Natural timber was selected as being better for the health of the children, rather than laminate or other industrial wood products, which contain strong bonding agents.

The main corridor, which links the functional spaces, has enough variety to act as an ad hoc play zone, while the roof reaches down to the playground to provide shaded external areas. The upper floor, positioned at the end of the building where the roof rises to its highest point, contains meeting rooms and storage space. Each nursery room offers direct access to the playground, as does the staff room. In total, the structure provides 1232 square metres (13,260 sq. ft) of usable floor space.

This building demonstrates that a fairly ordinary brief does not have to produce an ordinary building. Instead, Endo has responded by designing a capacious form with exposed engineering to stimulate the imagination of the children.

OPPOSITE, LEFT: The roof structure provides a sheltered arcade. OPPOSITE, RIGHT: The two-storey element contains office space on the first floor. CLOCKWISE FROM TOP LEFT The school is in a district of little architectural interest; the roof forms an entrance canopy; spaces flow through the building, creating informal play areas; the library; the semi-geodesic structure is formed from flat panels and steel props.

Fuji Kindergarten

Tachikawa, Tokyo

TEZUKA ARCHITECTS, 2005–2007

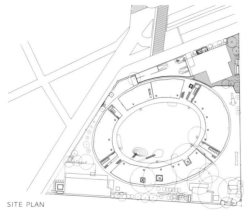

SITE PLAN

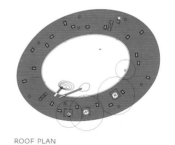

ROOF PLAN

BELOW, LEFT The roof is part of the play space. BELOW, RIGHT Rooftop access. BOTTOM Aerial view. The building's organic form is derived from a hand-drawn circle.

The Fuji Kindergarten has won a flush of awards, including the prestigious Architectural Institute of Japan Prize in 2008, but it has humble origins. When Tezuka Architects visited the kindergarten previously on the site, they discovered a place that badly needed a facelift but was nevertheless successful in ensuring the happiness of the children. The simple, bungalow-like linear structure stood in a large garden filled with old zelkova trees. The husband-and-wife team therefore tried to preserve some of the informal openness of the original building in their bold design for the new scheme.

The building is immediately distinguished by its oval structure, a shape, Tezuka says, that is loosely based on a hand-drawn circle. The irregularity is felt physically rather than visually. As nature was an important element of the architectural programme for this building, the architects carefully constructed foundations that would not interfere with the tree roots. The trees have therefore become part of the life of the building; indeed, the fact that everyone is free to roam over the roof brings the children into close contact with them.

To preserve the children's adventurous spirit, both the client and the architects at first considered ringing the eaves with safety nets instead of putting up railings along the edges of the roof. The planning authority vetoed this idea, and a compromise was reached whereby the eaves (just 2.1 metres/6⅞ ft from the ground) become benches from which children can observe the goings-on below; railings are installed but their balusters are wide enough to let children's legs through, so that they can dangle their feet over the wide gutter. Safety nets still feature, however, covering the gaps

around the trees on the roof. Children are allowed to jump into and nestle in the nets under the ancient trees.

Classrooms are divided by removable furniture and storage boxes rather than fixed walls. The only enclosed rooms are the toilets; the rest of the building operates more or less as a continuous space with no dead-ends. Sliding doors can be stacked one in front of the other, almost disappearing and blurring the boundary between the indoor and outdoor spaces. The Japanese tradition of taking off one's shoes before entering a building keeps the interior just about mud-free.

'Modern conveniences have deprived children of sensation', Takaharu Tezuka says. 'They don't know that when it rains, the soil gets wet. They don't know that if a person is hit, they get injured. They don't know the reason why a light bulb glows. What we want to teach through this building is common sense.'

LEFT The projecting roof extends the classroom space.
BELOW, LEFT Trees were left to form part of the interior spaces.
BELOW, RIGHT Furniture is simple and flexible; walls do not reach the ceiling.

Fukusaki Hanging Garden

Osaka City, Osaka Prefecture

KENGO KUMA & ASSOCIATES, 2002–2005

BELOW The building is backed by the perimeter fence of a golf driving range. OPPOSITE, LEFT Orange vinyl strips provide an ethereal quality. The building can be taken apart by simply unscrewing the bolts. OPPOSITE, RIGHT The transparent façade means that the building changes considerably in appearance between day and night.

Set in an industrial area of Osaka City, the Fukusaki Hanging Garden is a multi-purpose event hall for local schoolchildren. Envisaged as a temporary structure (to be taken down and moved elsewhere in ten years' time), it is also an investigation into what the architects call 'weak' architecture, a tangible polemic against buildings made entirely of concrete, which are authoritarian, monolithic and inflexible.

The building takes its cue from the local vernacular of industry and distribution (note, however, a particular influence the nearby driving range may have had on its structure, for there is an ethereal quality about both), and makes it fun, even beautiful. Constructed from industrial slabs – each consisting of two ultra-thin steel sheets clamped together by high-tension bolts and holding an iron grid between them – the whole building can be taken apart by simply unscrewing the bolts. The orange glow is derived from hundreds of translucent vinyl strips (of the sort that hang over industrial refrigerator openings or the doors of warehouses) suspended from the steel frame.

Everything about this building – including the bitumen floor – is industrial, and, although christened 'hanging garden', it is hardly garden-like. However, the vinyl strips, which take the place of walls, doors and windows, turn it into a place of exploration for children, who can roam freely and push through to new spaces wherever it suits them, as though they were in a wood – or indeed a garden. Walls, so the concept goes, are psychological as well as spatial barriers, so these utilitarian strips are offered as a gesture of flexibility, freedom and playfulness.

The two-storey building is accessed on the ground floor through an opening made by the absence of one of the slabs on the façade, and on the first via a gentle ramp that cuts diagonally across the façade. The building is artfully composed, especially its principal elevation of light and dark rectangles, sliced by the bright diagonal. Ostensibly a robust, no-nonsense container, it repays close examination, bearing out the postmodernist idea that industrial materials can be re-appropriated to softer, more social uses. It is the children themselves, however, who ultimately benefit from the permeable screens – postmodern or not.

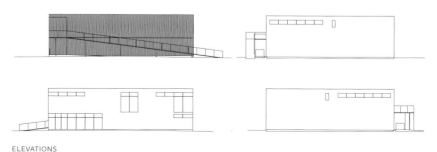

ELEVATIONS

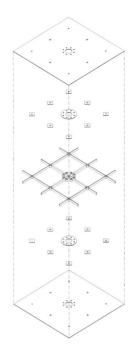
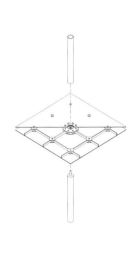

FLOOR DETAIL

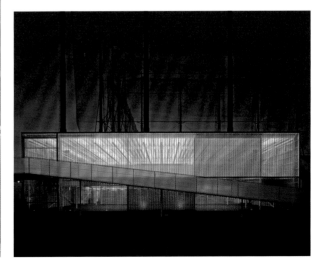

EDUCATION 139

Fukutake Hall, Tokyo University, Hongo Campus

Bunkyo Ward, Tokyo

TADAO ANDO ARCHITECT & ASSOCIATES, 2005–2008

Fukutake Hall, completed to mark Tokyo University's 130th anniversary, was designed as a new type of learning facility. The building provides space for researchers and postgraduate students to collaborate in a cross-disciplinary atmosphere. What makes the building particularly distinctive is its proportions: it is 100 metres (328 ft) long but just 15 metres (49 ft) wide. Architect Tadao Ando says the façade was modelled on the Sanjusangendo Temple in Kyoto, famous for the 1000 statues of Kannon (Avalokitesvara), Buddhist Goddess of Mercy, that line its long timber hallway. In fact, Fukutake Hall's regularly spaced columns, overhanging eaves and vertical grilles all echo the temple vernacular.

Ando also responds to a line of mature camphor trees, which forms a 'green belt' between the university and a busy road. He originally considered placing the entire building underground, but the amount of buried pipework under the Hongo Road prevented that plan. After a number of iterations a compromise was found by building two storeys below ground and two above.

Fukutake Hall faces the university campus, but visitors get only a glimpse of it at first through gaps in a solid concrete wall that runs the length of the building in front of it. The wall also acts as a buffer between the mix of old and new campus buildings, which have evolved over time, and Ando's newest addition. Past this shield, however, a surprise is in store: a dramatic precipice hollowed out by two giant, symmetrical external stairways reaches down to the underground chambers.

Lifts, offering another way down to the 200-seat lecture theatre at the bottom, divide the building into two halves. One contains a series of rooms for staff and students, and a cafe. The other unfurls a large open-plan area for research and study. Such recent projects as Shibuya Station (pages 52–53) and Omotesando Hills Shopping Centre prove Ando's adeptness at creating efficient circulation to deal with large crowds of people. For him, it seems, it is as easy as drawing a circle on a piece of paper. Unlike his commercial buildings, however, Fukutake Hall has the simplicity of plan and resolution of a pavilion – only this pavilion reduces its presence by burrowing into the ground.

OPPOSITE The hall provides an elegant perimeter to the campus. The bulk of the building is underground. RIGHT The roof projects to form a canopy. BELOW, FROM LEFT: The building provides a modern counterpoint to early twentieth-century architecture; external stairway to the lower levels; other campus buildings can be seen through a slit in the wall. BOTTOM The 200-seat lecture theatre and the open-plan research and study area.

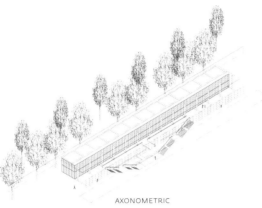
AXONOMETRIC

SECTION

EDUCATION 141

Future University – Hakodate

Hakodate, Hokkaido

RIKEN YAMAMOTO & FIELD SHOP, 1998–2000

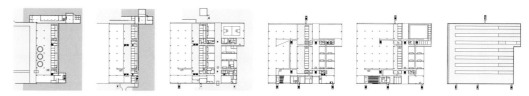

(LEFT TO RIGHT) GROUND- TO FOURTH-FLOOR PLANS; ROOF PLAN

Built on a sloping site with a panoramic view of the city of Hakodate and the sea, this large university building, designed as an all-enveloping enclosure, was partly inspired by the university's motto that open space engenders an open mind. The Future University of Hakodate, which teaches systems information science, accommodates two smaller departments, each with the focus on computing: complex systems and media architecture.

All the facilities necessary for advanced research are found in this monolithic building: teaching spaces, research laboratories, offices, a library and an auditorium. Clad in square panels of glass, it can be accessed either through the main entrance, which opens into a large atrium, or through the back entrance, at the end of a colossal stepped space called 'the studio'.

The atrium, encompassing five storeys, is busy with steel rods, steel railings and steel mesh supporting bridges, lift shafts and stairways; the sight line, although somewhat pulverized, is preserved. Workshops, lecture rooms, offices and libraries, all visible (but, crucially, not audible) from this area through the transparent dividing walls, line both sides.

The studio, at the other end of the atrium, is a larger-than-life open space stretching the width and height of the building. The four levels are arranged in a stepped structure, supported at regular intervals by solid concrete columns. Accessed via lifts and bridges, the studio was designed to encourage interaction among students in all departments. The laboratories for teachers and researchers, the student canteen and the shop are placed around the edges, partitioned again by glass screens.

Visibility is fundamental to this building; its design is thus uncompromisingly modern. Plenty of concrete, steel and glass have been used to create a space that theoretically encourages teamwork (open-plan areas) as well as solitary thinking (flexible, modular-based units) and inspiration (transparency). A new annexe designed by the same architect has recently been built, but it lacks the 'wow factor' still shown by the original building.

SECTION

OPPOSITE The glazed façade flanking the studio. BELOW Interiors are characterized by both formality and informality. BOTTOM A robot on display. RIGHT, TOP The studio is a colossal cascading space designed to encourage interaction between students in all departments. RIGHT, BOTTOM The five-storey atrium provides a clue to the scale of the building.

EDUCATION 143

Gifu Academy of Forest Science and Culture

Mino, Gifu Prefecture

ATSUSHI KITAGAWARA ARCHITECTS, 1998–2001

BELOW The academy was conceived as an 'inhabited hill'. BOTTOM These workshops straddle traditional and contemporary design, with the historical gridded *mengoshi* walls a repeating feature. OPPOSITE Cedar logs are joined in tree-like structural forms.

The Gifu Academy of Forest Science and Culture is a specialist centre of excellence for all matters relating to timber and forestry, including woodland management, timber architecture and furniture manufacture. Designed by Atsushi Kitagawara, who rose to prominence in the early 1980s with his sculptural approach to architecture, this educational facility can be described as an 'inhabited hill' (*satoyama*), intended to foster the 'co-existence of man and the forest'.

The architectural expression of this search for coexistence lies in the use of traditional Japanese carpentry methods for the timber buildings spread across the site; nails and other metal hardware are kept to a minimum. The buildings at Gifu Academy are a combination of contemporary design and age-old techniques of jointing wood. Also, the timber was sourced on site by removing trees for the general good of the forest (a standard practice that provides more space and light for the remaining trees).

The most obvious reference to this close relationship between inhabitant and forest is, perhaps, the tree-like structural supports employed in part of the estate; cedar logs are carefully joined in a branching format, which helps in the creation of large spans. Other architectural devices include the large-format *mengoshi* walls and columns, where the structure is expressed as a rigid grid (without ironmongery, of course). This form admits filtered light, as a deliberate reminder of the shadows and darkness found in the forest itself. According to Kitagawara, 'every piece of wood preserves a forest and its "darkness" within it'.

Gifu Academy demonstrates a huge variety of ways in which wood can be manipulated. Sometimes it appears as long beams, at others as trusses. Other materials also feature, such as the soft elastic fabrics stretched across the wooden beams, displaying Kitagawara's characteristic whimsical touch. Providing 7700 square metres (82,880 sq. ft) of space on 8 hectares (19¾ acres) of forest, the collection of buildings comprises everything from accommodation and workshops to machinery sheds and an outdoor theatre.

EDUCATION 145

Gunma Kokusai Academy
Ota, Gunma Prefecture

KAZUHIRO KOJIMA + SUSUMU UNO +
KAZUKO AKAMATSU/CAN + CAT, 2003–2005

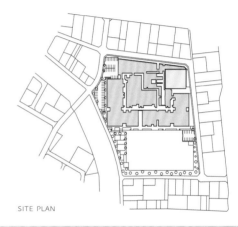

SITE PLAN

This school on the edge of the city sprawls over 2 hectares (nearly 5 acres) of land and provides roughly 8000 square metres (86,000 sq. ft) of floor space. As well as being generous with space, it offers an unusual teaching system. All subjects are taught in English, apart from Japanese and social studies classes. This 'immersion' programme is designed to envelop students in English so completely that the language comes naturally. The architects felt that this dramatic departure from standard teaching practice required a different kind of architecture: rather than lining up classrooms for the discipline of intensive instruction (as is typical in Japan), they envisaged a programme that would encourage interaction.

The result is a complex of buildings designed to mitigate authoritarianism and to promote feelings of homeliness. Clad in metal plates, the low-level buildings, which broadly conform to the massing of the residential area in which the school sits, are situated on streets and around courtyards like a town within a town. Individual 'houses' serving particular age groups are linked, while such specialist facilities as workshops and a media centre are in their own separate buildings. Covered walkways connect the various elements into a unified whole.

The use of timber throughout the development further softens the institutional nature of the school, adding a layer of domesticity to the buildings. Even the metal plates on the façades are layered like strips of timber, giving the impression more of a mountain hut than of a school. Wood – chipped, woven and intersected – is key to the internal spaces, which are arranged for transparency and to provide long views. Double-height learning areas, flooded with light from clerestory windows, are flexible enough for group- and project-based work.

C+A (Coelacanth and Associates; CAt signifies the Tokyo office and CAn the Nagoya office) takes the 'fluidity and instability' of life as a starting point for architecture, and seeks to create a building around the ways in which spaces are used. Form and façade treatment are therefore secondary to what goes on in the buildings. The practice has recently undertaken work in central and southern Asia, but this Japanese project, smaller in scale than its other buildings around the world, represents its core values very closely.

FLOOR PLAN

OPPOSITE: The campus is designed along village lines, almost like a residential development. The site is characterized by streets and squares. TOP LEFT AND BOTTOM LEFT Individual 'houses' serve different age groups, and are linked by covered walkways into a unified whole. TOP RIGHT AND BOTTOM, CENTRE AND RIGHT Timber is used inside to promote a feeling of homeliness.

EDUCATION 147

Ina-Higashi Elementary School

Ina, Nagano Prefecture

MIKAN, 2003–2009

EAST ELEVATION

WEST ELEVATION

LEFT Exterior view. BELOW AND OPPOSITE, BOTTOM A freestanding library 'pod'. OPPOSITE, TOP Multi-purpose space on the upper level extends the classrooms.

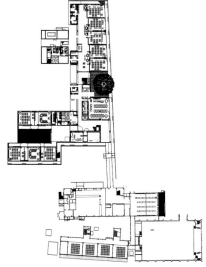

GROUND-FLOOR PLAN

School closures are common throughout Japan as the country's population ages and its birth rate drops. But thanks to an influx of Japanese immigrants from Brazil, Ina – a community of 60,000 – has been blissfully immune to such problems. With classrooms filled to capacity, the city nominated the Ina-Higashi Elementary School for a sizable addition and then held a competition to find a suitable designer. A fully developed building plan was not required; on the contrary, the competition organizers sought a firm willing to generate discussion among and incorporate input from teachers and parents as well as bureaucrats. Good listeners and good designers, Mikan took the prize.

For the first three years of the project the Tokyo-based architects travelled regularly to Ina, to brainstorm with the community and conduct workshops to identify how best to blend their new building with the old and allocate interior space within the bounds of the established curriculum. Out of this prolonged design phase came a modest L-shaped facility that comfortably reconciles a variety of existing conditions: the four buildings still in use, the playground, an adjacent house acquired for extra-curricular activities, and a magnificent cherry tree, the pink blossom of which heralds the start of the new academic year each spring.

The school's main entrance is at the intersection of its two legs, next to the administrative offices. From here circulation conduits branch off in two directions, widening into multi-purpose spaces that act as extensions to the classrooms. Upstairs, the library connects the two wings. Because of their proximity to the outdoor play areas, the downstairs classrooms belong to the first- and second-year students; sixth-year pupils study upstairs, and the third, fourth and fifth years are housed in one of the older buildings next door.

Reinforced-concrete load-bearing walls organize the ground floor, which is finished warmly in wood. But upstairs, where a steel frame supports the undulating ceiling grid, rooms flow easily from one to another. Instead of full-height solid walls for separation, clear glass inserts mitigate the fluctuations overhead and extend sight lines. Clad in painted plasterboard, the wavy grid dips downwards in the library, sheltering a quiet spot to curl up with a book, but soars again in the music room, where the noise level crescendos.

Outside, a pitched roof caps the building and prevents snow from accumulating, since Ina sits in a valley in the Japanese Alps. Instead of challenging the scenery, the school's exterior works with it. While its indented, irregular perimeter embraces greenery and gardens, the grey of the concrete blends effortlessly with the background.

EDUCATION 149

Kaito Workshop
Atsugi, Kanagawa Prefecture

JUNYA ISHIGAMI + ASSOCIATES, 2005–2008

At Kaito Workshop, about an hour's train ride west of Tokyo, students at the Kanagawa Institute of Technology are encouraged to make whatever they wish, outside the constraints of the curriculum. An intricate network of steel pillars – not easily visualized from the building plan, with its amalgamation of dots – is judiciously positioned to support and enhance the glass structure. The pillars not only are load-bearing but also serve as partitions, and form a vague path that allows transport vehicles to travel through from one end of the building to the other. In white, and facing different directions, heavy-duty air-conditioning units also form part of the partitioning, and keep the glass building at an agreeable temperature up to a certain height (the air near the ceiling reaches 40°C/104°F in the summer). Otherwise, the workshop is open plan, allowing people to roam freely.

The building's concrete base is lifted off the surrounding bitumen just enough to give the anti-gravitational effect. Glass panels 5 by 1.5 metres (16⅜ by 4⅞ ft) are fixed seamlessly together to work as a curtain wall and showcase the students' work.

This project could be viewed as a reworking of Koga Park Cafe (1998) and Naoshima Ferry Terminal (2006; pages 48–49), both designed by SANAA (Ishigami worked for Kazuyo Sejima for a few years before setting up his own practice). Although the two firms share a number of interests, on closer inspection we discover that Ishigami's focus is rather different. The mirrored partitions and poles were effectively used in SANAA's buildings to promote transparency, but Kaito Workshop's pillars, which are excessively dispersed, defy such lucidity.

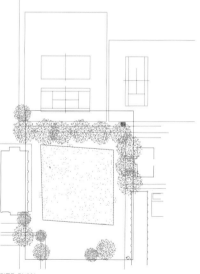

SITE PLAN

For now, it seems, practicality is set aside for playfulness. Delicate steel chairs and tables – which Ishigami designed for this project – radiate a serene, almost otherworldly air. This is in sharp contrast to other areas, which are weighed down by heavy machinery. In this new space, Ishigami's very first built project, art and architecture converge effortlessly.

THIS PAGE An intricate network of steel pillars – represented by dots on the site plan – supports the delicate roof. OPPOSITE Steel pillars not only bear the load but also serve to some extent as partitions. Ishigami selected various types of indoor plants as part of the interior furnishing, and desks and chairs were salvaged from the dismantled campus building.

EDUCATION 151

Minami-Yamashiro Primary School

Minami-Yamashiro, Kyoto Prefecture

RICHARD ROGERS PARTNERSHIP, 1995–2003

While the houses and shops in the sleepy village of Minami-Yamashiro huddle beside the winding main road at the foot of the valley, this new primary school by Richard Rogers Partnership (now Rogers Stirk Harbour + Partners) stands in a prominent location over the hill, its stainless-steel roof peering over the green horizon.

The brief was ambitious: to reverse the dwindling rural demographic trend through architecture. Four local primary schools, a centre for adult education, a nursery and a welfare centre for the elderly were all to be accommodated in the new masterplan, which was to become the hub of the community (although eventually both the nursery and the welfare centre were designed by other firms). Such a space, it was hoped, would reinvigorate the local population by symbolizing a new, thriving identity.

This was a rare chance for Rogers, as an urban enthusiast, to dwell on the quieter side of life and design a building in the countryside. The huge expanse of land allowed Rogers and his Tokyo office plenty of space to plan the school's main building, a gymnasium, an open-air swimming pool and a car park. It all had to be done, however, in the most economical way possible, with a budget of only £11.8 million.

The practice came up with a simple modular system, the components of which could be extended endlessly to provide an open and flexible space. The repetitive grids of 8 square metres (86 sq. ft) are framed in concrete and clad alternately in glazing panels with aluminium borders (reminiscent of traditional lightweight *shoji* paper doors) and solid concrete painted in vivid colours. These colours differentiate areas and functions, and they make a sharp contrast with the subdued, skin-toned cladding on most municipal schools across Japan.

The internal layout focuses on the triple-height concourse, vibrantly coloured as if it were a deep sea. Bridges across it allow pupils direct access to the playground on the intermediate level. The hallway and classrooms are evenly illuminated by light falling through the gaps between the walls and the sequentially pitched rooftop, which is elevated on steel rods. In terms of scale, this is a 'mega-structure' with an urban efficiency. The project, however, fits easily into its rural environment because of its grid system – which offers smaller, more manageable spaces – and various fixtures, such as sliding doors, all of which are sympathetic to classic Japanese construction methods. Recognizing its elegance, the Royal Institute of British Architects granted the school an award in 2004.

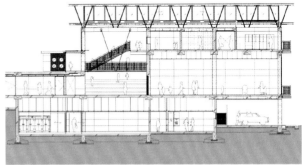

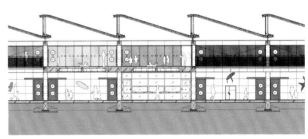

SECTIONS

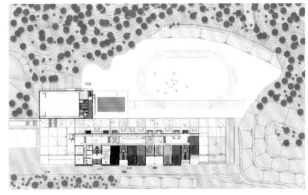

SECOND-FLOOR PLAN

OPPOSITE Unlike those of other school buildings in Japan, these façades are adorned with bright colours. The swimming pool and gymnasium are seen top and bottom. LEFT The triple-height concourse in the main building. LEFT, BOTTOM A sequentially pitched roof allows natural light to flood in.

EDUCATION 153

Mode Gakuen Cocoon Tower
Shinjuku, Tokyo

TANGE ASSOCIATES, 2006–2008

BELOW, LEFT Cocoon Tower's fifty-storey ellipse contains a trio of vocational schools. BELOW, CENTRE The tower stands out in Shinjuku's chaotic urban environment. BELOW, RIGHT, AND OPPOSITE, TOP Like silk thread, a diagonal steel frame cross-hatches the building's skin. OPPOSITE, BOTTOM LEFT Internal stairs connect blocks of classrooms. OPPOSITE, BOTTOM RIGHT The 'eye' of the tower offers an outstanding view of Tokyo. OVERLEAF A classroom in which students can devise new Nintendo DS games.

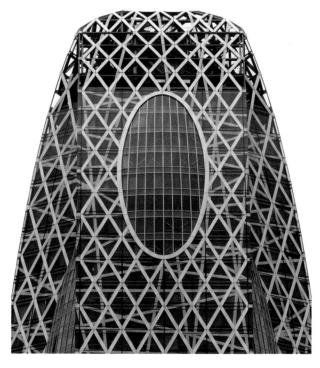

No area reflects Tokyo's density and visual incoherence better than Shinjuku. One of several commercial centres in the city, it is also a major transport hub, serving more than 3 million commuters each day. An intricate web of neon-clad low-rise structures elbow one another on one side of the station, and orderly blocks of skyscrapers stand stoically shoulder-to-shoulder on the other; between these two extremes few individual buildings stand out. But Mode Gakuen Cocoon Tower, a statuesque fifty-storey ellipse with a striking diagonal cross-hatch pattern, is an absolute head-turner.

Although it hardly looks the part, Cocoon Tower is a vocational college with 10,000 students. Divided horizontally, the high-rise campus contains three separate divisions: fashion, computer technology and healthcare. All are administered by a parent company, Mode Gakuen, which was founded by two brothers who inherited a small sewing school from their mother. After turning her modest venture into a successful fashion institute, they set their sights on other areas of growth. Intending to consolidate their holdings under one roof, the brothers held a design competition and awarded the commission for a new building to Tange Associates.

The task facing the architects was to create a technical school the likes of which had never been seen before. They met this challenge with a two-part scheme consisting of an 'egg' and a 'cocoon', joined at the base. Hugging the ground, the squat ovoid holds two lecture halls. The soaring, elongated ellipse contains three vertical stacks of classrooms evenly spaced around a lift–stair core and interspersed with triple-height atriums. With spectacular views, they function like playgrounds where students can relax and socialize. As the tower tapers, the atriums are replaced by lounges, which fill out the floor plate at the top. Designed to draw people in, the building curves inwards at the bottom and is surrounded by landscaped terraces. In addition to entrances on either side at street level, an underground pedestrian passageway links the building directly to the station.

Cocoon Tower's flashy appearance contrasts sharply with the surrounding staid corporate headquarters as well as Tange's own City Hall near by. But its unique exterior is more than cosmetic. Structure and skin combined, it consists of double-layered glass and a diagonal steel frame that acts like a hard shell to support the building. A second pattern of strips of white dots applied to the glass cuts down on glare. Together the overlapping white lines read as threads wound around a cocoon.

Aptly named for its function, the building with its bulging form incubates budding worker bees. But Cocoon Tower's sophisticated appearance enables them to study in style.

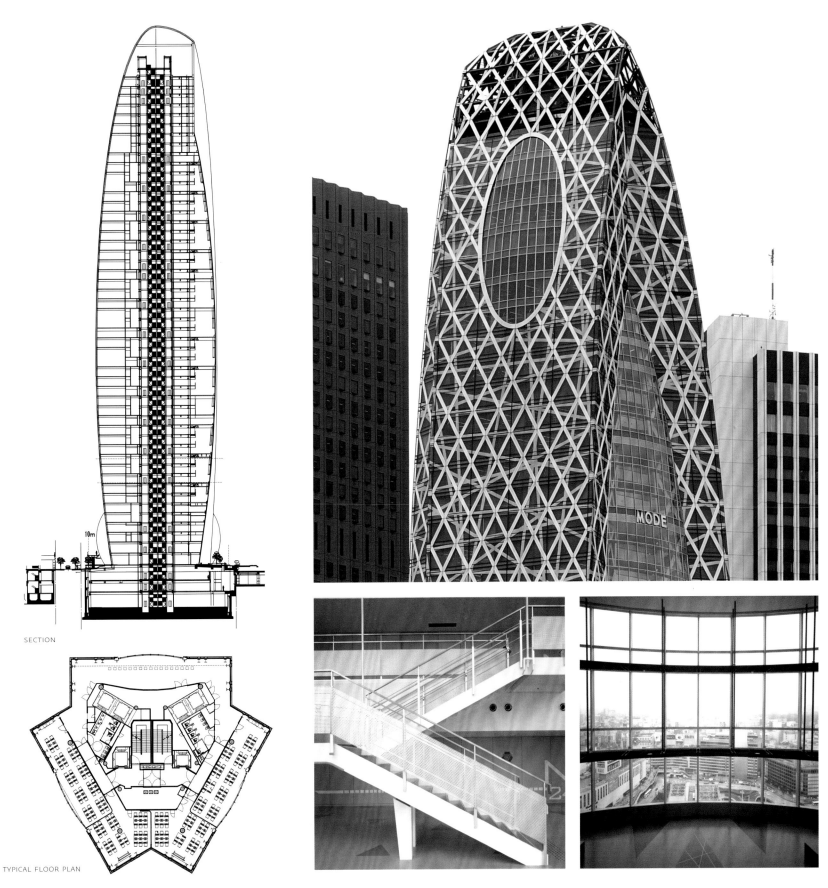

SECTION

TYPICAL FLOOR PLAN

EDUCATION 155

Mode Gakuen Spiral Towers

Nagoya, Aichi Prefecture

NIKKEN SEKKEI, 2004–2008

ELEVATIONS

Spiral Towers, in the very centre of the city of Nagoya, is the first of two buildings to be completed for the renowned fashion college Mode Gakuen. Nikken Sekkei, a large firm with nearly two thousand staff, won a closed competition to design this building, while Tange Associates won the commission to design Mode Gakuen Cocoon Tower in Tokyo (pages 154–57).

Topped with a helipad, the thirty-six-storey building with three towers twisting up and around its central core is an eye-catcher, and crowds gather to view it from the nearby Marriott Hotel's fifteenth-floor reception hall, above the city's main railway station. Such a complex form was devised to accommodate three wholly separate vocational colleges in one place: Nagoya Mode Gakuen, which offers courses in fashion and design; HAL Nagoya College of Technology & Design, a specialist computing and technology college; and Nagoya Isen, an establishment allied to the therapeutic medical profession, teaching such skills as 'medical make-up'.

Although a serious structural achievement, especially when one considers the seismic conditions the building has been designed to withstand, the tower is in fact a simpler form than it first appears. It is composed of a stiff vertical core of concrete-filled steel columns, to which the floor plates are attached, each one carved away diagonally to give the building its distinctive corkscrew appearance. The glazing system emphasizes the twist, and protective grids over the glass carry the spiral movement up into the sky.

The original idea of accommodating three colleges in the separate wings of the building, however, remains firmly conceptual. During the design process, Nikken Sekkei discovered that in order to carry through such an idea, the access points would need to be arranged separately – not a feasible solution either structurally or economically. All the lifts therefore had to be incorporated into the central core. Consequently, the interior layout is rather more conventional than one might expect; classrooms for each college are dispersed on different floors at random, their oval plans the only reminder of the intriguing form that begot them.

Despite such a disappointment, Spiral Towers is still remarkable, as it incorporates a series of sustainability measures, including a double-skin façade. The façades of the three wings comprise two layers of glass: a vertical inner layer and a sloping outer layer. The space between forms an environmental buffer zone, allowing air to circulate naturally via window vents, even at very high levels. Blinds, to prevent glare, are fitted within this cavity. With such environmental credentials, the building is certainly more than just a striking high-rise.

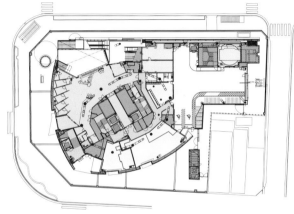

SITE PLAN

OPPOSITE, LEFT The position of signage emphasizes the curvilinear form. OPPOSITE, RIGHT The structure is composed of three elements, which wrap around a central core. BELOW Three-dimensional computer modelling was crucial in realizing the complex design. RIGHT, TOP Reception area. RIGHT, CENTRE A double-skin façade helps provide environmental control. RIGHT, BOTTOM A lecture theatre.

EDUCATION 159

Musashino Art University Library
Kodaira, Tokyo

SOU FUJIMOTO ARCHITECTS, 2007–2010

The new library of Musashino Art University does not simply house bookshelves – it *is* a bookshelf. Some 550 metres (1800 ft) long, it consists of a colossal spiralling wall of books contained in a box of clear glass. The wall begins at the library's main entrance and terminates in the circulation desk at the heart of the building. In between it yields a labyrinthine sequence of spaces where browsers can lose themselves in the 300,000-volume collection.

The new building, on the outskirts of Tokyo, belongs to an art school of five thousand students, and is the product of architect Sou Fujimoto, who won the commission in an invited competition. The university wanted to upgrade its existing library and gallery building, erected when the campus was first developed in the 1960s. Its aim in hiring Fujimoto was to refurbish the old building as a gallery and attach a new library to it.

At first, Fujimoto envisioned the library as a 'book forest', where students could wander and relish chance encounters with the collection's beautiful art books. As the project began to take shape, Fujimoto realized it needed a form that addressed not only the activity of unguided exploration but also that of specific investigation. His answer was the building's unique spiral form: its meandering spaces provide opportunities for unexpected discoveries, while its recognizable, linear path facilitates the organized quest for information.

Users enter the two-storey building through either the main entrance on the ground floor or the secondary entrance upstairs, which links the new library to the old building. The campus is laid out like a chequerboard, its various buildings interspersed with open spaces. Fujimoto's new library occupies a central and heavily trafficked spot near the university's main cafeteria, filling in one of the remaining voids stipulated in the original masterplan.

To mitigate this filling-in, Fujimoto strove for light and transparent architecture. While there was little he could do about the solidity of the books, he left plenty of openings in the plywood shelving and held everything together with an unobtrusive steel frame. Enriching the sense of depth, the apertures create cross-views inside as well as from the outside.

In contrast to the concrete buildings that anchor the campus, Fujimoto's glass box practically melts away, exposing its inner workings to everyone who passes, and straddling the line between the solids and voids of the campus.

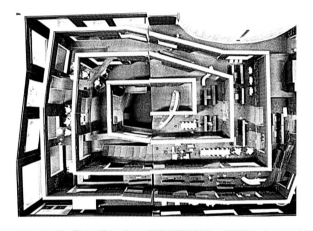  

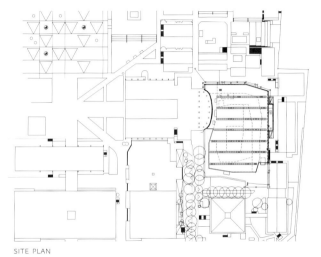 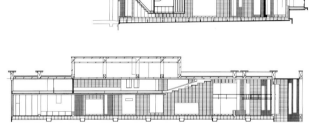

SITE PLAN

SECTIONS

OPPOSITE Interior elevation. TOP, LEFT AND RIGHT A 550-metre-long (1800 ft) spiralling wall lined with bookshelves defines the interior. TOP, CENTRE Conceptual diagram. ABOVE Two views of the reading room.

Rikuryo Alumni Hall, Kitano High School

Osaka City, Osaka Prefecture

AMORPHE, 1998–2004

BELOW The Alumni Hall was paid for by charitable donations. OPPOSITE, TOP LEFT The building is composed of a plinth and a cantilevered curve. OPPOSITE, TOP RIGHT AND BOTTOM The sweeping curve of the exterior continues inside.

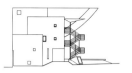
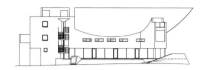

ELEVATIONS

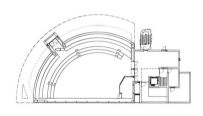
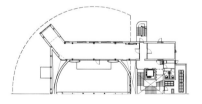
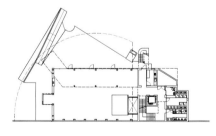

PLANS

Alumni of Kitano High School include the architect Kiyoshi Sey Takeyama of Amorphe, as well as Osamu Tezuka, the creator of the classic animated character 'Atomic Boy'. Paid for through charitable donations, the alumni hall was completed in 2004 after a design period of six years.

The building was intended to provide a 'place of memory' and to mediate between the school and the wider world. Containing a library, an office with a separate meeting room, an event space, an exhibition room and a performance hall, the building provides an iconic presence for the school, which is positioned on a busy road in Osaka's Yodogawa District.

Built from steel and reinforced concrete, the hall is composed of two principal elements: a three-storey plinth (including basement); and a curving form containing the performance hall, which floats above and adjacent to the more straightforward linear spaces that serve it. This composition is not as wilful or arbitrary as it might at first appear. The curving element is geometrically part of a notional sphere, which, left to form in the mind of the viewer, represents the possibilities of graduation: the potential of the future and the idea that the school can do no more than prepare a person for life in the wider world. The presence of a sphere also hints at the Earth and, by extension, the Universe.

Directly countering the crispness of these pure geometric forms is a black volume that appears to support the sphere's underbelly against the will of its own weight. Cast with expandable metal grids, which give a rough finish, the bulging form actually seems to cave in, so black that it sucks in everything, even its own shadow, like a black hole. A cantilevered structure leans out from it, as if defiantly resisting the force of gravity.

This underbelly then sweeps into the ground-floor event room as a curvy black metal ceiling. The seating in the performance hall is also dramatically tiered, echoing its spherical shape. The basement holds a small gallery space as well as a secure storage area: a useful facility, since the combined work of the alumni (at least those pieces that can be hung on the wall) is worth billions of yen. In more earthly terms, the building functions as a 'place of interchange', where ideas can be shared and events hosted in the manner of a club.

EDUCATION 163

Tama Art University Library
Hachioji, Tokyo

TOYO ITO & ASSOCIATES, ARCHITECTS, 2004–2007

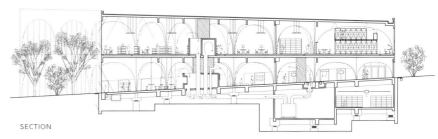

SECTION

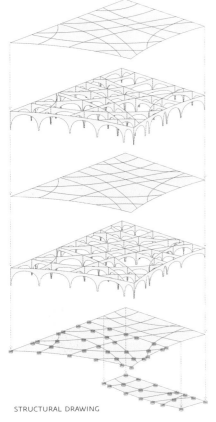

STRUCTURAL DRAWING

Commissioned by Tama Art University, where Toyo Ito is a guest professor, the library building demonstrates the same lightness of touch, humour and boldness that can be seen in his other works. As before, these qualities are expressed through a fascinating geometric composition of smoothly poured concrete and glass, but while some of his projects never seem to resolve entirely the disparity between interior and exterior, Tama Art University Library displays a new integrity.

The library is prominently situated next to the university's main entrance. Its exposed-concrete façade responds well to the surrounding campus buildings, most of which are clad in the same material. Providing a contrast to them, however, Ito combats the stiffness and heaviness of concrete by carving out great chunks to make elegant voids, inspired (so he says) by the nimble arches of women's high-heeled shoes. This inspiration suggests both the humour for which Ito is known and the great emphasis he places on lightness and the impression of movement. He moulds the north- and west-facing exterior walls out of gentle concave curves, moreover, allowing them to meet at sharp corners with a sweeping angularity more common to Gothic or Baroque buildings. These corners subtly amplify the mobility of the building and showcase the exquisite engineering feat it represents.

Inside, the leaping arches are repeated throughout the ground and first floors. The curves reverberate through the desks and bookshelves, which were designed specially by Kazuko Fujie, a fellow professor at the university. Perfectly at ease in this setting, students appear to linger as if they are reading books or watching DVDs under the branches of a great tree. Round discs disguising lights descend upon them to complete the circle of the arches. Soft, translucent curtains – not blinds – protect books from strong sunlight and tone down some of the harshness generated by the exposed concrete within.

The section of the library called the Arcade Gallery is open to the public. The gallery connects the south and north entrances, allowing the visitors who pass through it to experience the gentle slope that follows the topographical incline. Here, people are free to rest and possibly ponder why more educational buildings are not as delightful as this.

BELOW The walls meet at sharp corners with a sweeping angularity. OPPOSITE, CLOCKWISE FROM TOP LEFT Detail of the metal shelves installed under one of the arches; the customized bookshelves and round discs disguising lights all respond to the leaping arches; some arches span a 16-metre (52½-ft) void; the floor of the Arcade Gallery gently slopes, following the topographical incline; detail of the smooth, elegant façade; ceiling detail.

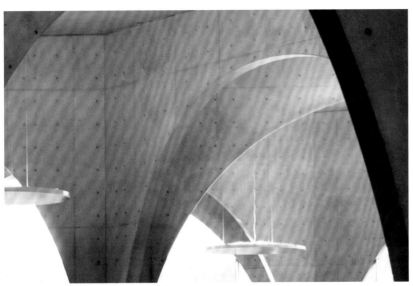

EDUCATION 165

Health and Religion

168 Baisouin Temple
172 Buddhist Temple
174 Chapel Aktis
176 Children's Centre for Psychiatric Rehabilitation
178 Leaf Chapel
180 Sendai Baptist Church
182 White Chapel

Baisouin Temple

Aoyama, Tokyo

KENGO KUMA & ASSOCIATES, 2001–2003

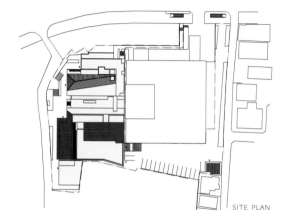

SITE PLAN

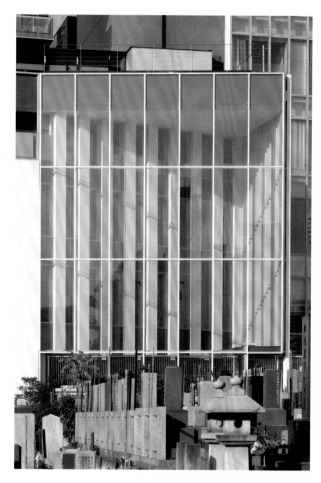

Occupying a site alongside both a century-old cemetery and a new fourteen-storey commercial tower, opposite Kengo Kuma's own office in the upmarket commercial district of Aoyama, the Jōdo Buddhist Baisouin Temple is part of a conglomeration of old and new. Kuma, to whom the area is very familiar, has attempted to make his building a public place – opening up to the city and its flow of people – rather than an isolated, monastic precinct. 'In the same way as temples and churches used to serve as urban public spaces, I planned to integrate the city and the temple into one space', he explains.

The complex is large: two basement levels and five storeys above ground, with a library and three principal halls, one of which can accommodate 350 people. The temple functions as a community facility, providing space for concerts, funerals and other large group events. It has become something of a cultural focal point in the district.

The distinctive louvred, metal-clad walls, providing an envelope for the concrete-and-steel structure beneath, are deeply ribbed, inspired by the roof tiles of the original temple, catching the light or being thrown into shadow in varying intensity. One, angled rather like a lean-to, conceals an entrance to the community halls at one end of the building. The darkness of this slanted wall is in stark contrast to the whiteness on the other faces of the building.

Inside, Kuma applies an entirely different treatment. Here, the light is softer and more ambient: a warm glow from the backlit fibreglass panels – which, reminiscent of traditional Japanese *shoji* screens, line the interior walls in the main hallway – offsets the cooler shade of the heavy stone floor. He excels in an architectural approach to interiors, and, often using such traditional materials as bamboo, manages to create modern palettes that work beautifully. 'The problem with Modernism is that it is still under the influence of Classicism, in respecting proportions and the beauty of shape. But for me that is a secondary thing', he says.

On the top floor, the 'water garden' outside the VIP waiting room reflects the Tokyo skyline, framing the iconic Tokyo Tower and Kohn Pedersen Fox Associates' new development, Roppongi Hills. In the cemetery below rest the ashes of the late Kisho Kurokawa, an outspoken advocate of Modernism – signifying perhaps the end of one era and the beginning of another.

ABOVE, LEFT The temple overlooks a century-old cemetery. ABOVE, RIGHT Principal entrance. OPPOSITE, CLOCKWISE FROM TOP RIGHT View over rooftop; the VIP waiting room; natural light (RIGHT) and translucent, backlit fibreglass panels (LEFT) gently illuminate the temple's interior; side entrance. OVERLEAF Outside the waiting room, Tokyo's magnificent skyline is reflected in the rooftop pool.

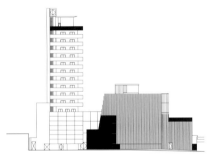
WEST ELEVATION

HEALTH AND RELIGION 169

Buddhist Temple

Kagoshima Prefecture

HEATHERWICK STUDIO, 2004–

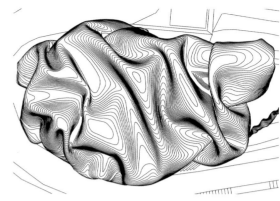

AERIAL DRAWING

This Buddhist temple, unbuilt at the time of writing, is a deliberately unconventional interpretation of the building type, and will, if built, contain an ossuary with space for the cremated remains of 4000 people. After an international search and an intensive period of consultation and interview, London-based designer Thomas Heatherwick was specially selected for being neither Japanese nor Buddhist, and therefore free from preconceptions. The client, a Shingon Buddhist priest, took him on a tour of Kyoto's shrines and temples, explaining that he was expected to understand these places but not mimic them. During a lengthy design process, Heatherwick began to explore the folding potential of fabric, partly inspired by the fact that the Buddha sits on a cushion. Using a scanning device at a London hospital, Heatherwick settled on one particular format of folds and had it preserved as a computer model.

The temple will be constructed of timber and glass, configured as a set of contours. Rather than creating a complex curve where the surface moves in three directions simultaneously, an undulating surface is achieved by placing one layer on top of another, culminating in a three-dimensional form in the way that magnetic resonance imaging creates a model of the human body slice by slice. Each contour represents a step, the upper surface of which is angled slightly downwards to facilitate rainwater run-off. Glazing is set into the structure like geological strata.

The hillside plot in Kagoshima, on the southern tip of Japan, is near the site of a famous battle in 1877, between the new government and the forces of the local samurai Saigo Takamori,

who died in the battle and to whom the temple is to be dedicated. The beauty of Heatherwick's solution to the brief is that its complex topography can be read as almost anything: landscape, the folds of the brain, an Escher-like composition of looped and infinite forms, or even a stack of history books. For a place of contemplation and soul-searching, this form is full of promise.

ABOVE Wooden model.
OPPOSITE Detailed views of the temple's folded form.

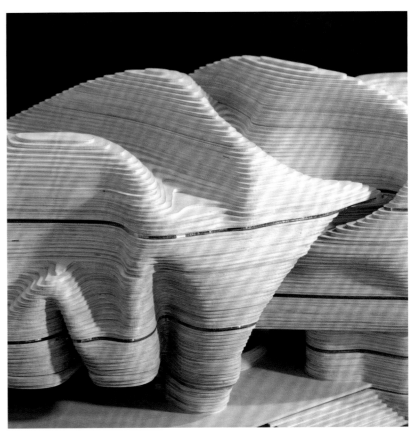

Chapel Aktis

Kyoto City, Kyoto Prefecture

AMORPHE, 2004–2005

The Kyoto-based practice Amorphe (principal: Kiyoshi Sey Takeyama) should know that the city in which it is based is not an easy place for architects. In the city centre, even McDonald's stores (before the franchise's recent makeover) must display their trademark in chocolate brown so as not to spoil the image of a city steeped in history. With its understated decorum, then, this wedding chapel in the grounds of the city's Brighton Hotel has a distinctive Kyoto feel.

Weddings are big business for large hotels in Japan, assuring them steady revenues even in a time of recession. Usually shrines and wedding chapels are set up inside the hotels, but when budgets allow the chapels can, despite – or perhaps because of – the lack of real faith, become the site for great experimental projects, such as Klein Dytham Architecture's Leaf Chapel (pages 178–79) and Jun Aoki & Associates' White Chapel (pages 182–83).

Clad in titanium zinc as if the metal were soft *furoshiki* (colourfully dyed cloth used by the Japanese to wrap and carry gifts), Chapel Aktis tapers into the sky in the shape of a small pyramid with its top cropped off. The diagonal pattern on its façade reinforces the idea of wrapping. Nearly windowless apart from a trapezoid dent of stained glass above the altar, and a few small holes and slits on the sides, it stands as a curious object, especially when viewed from the street.

The architects carefully choreographed the design to allow one-way progression only, so that there is no risk of different wedding parties accidentally meeting. A raised concrete walkway with a horizontal aperture beneath brings one party in, while the glass doors – framed by yet another

OPPOSITE The principal stair and exit. BELOW The chapel's form is asymmetrical. RIGHT, FROM TOP The principal stair and (to the right) the ceremonial route; the titanium zinc-clad façade; the interior of the chapel.

trapezoid – in the main chapel offer a way out for the previous party, with stairs acting as convenient stands for group photographs. Such an affair may sound highly regulated and disciplined, but Amorphe manages to suggest the sanctity of marriage with a distinct sense of procession and revelation. Inside the chapel an open space unfurls, with white walls and a high ceiling, minimally furnished with rows of slick black benches and a simple altar that echoes the non-linear nature of the building. The space seems to spread up and out, towards good fortune.

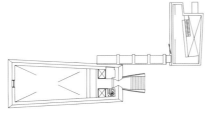
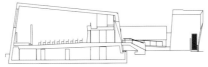

PLAN AND SECTION

HEALTH AND RELIGION 175

Children's Centre for Psychiatric Rehabilitation

Date, Hokkaido

SOU FUJIMOTO ARCHITECTS, 2004–2006

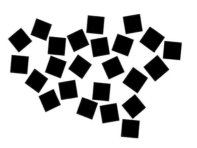

CONCEPTUAL DIAGRAM

GROUND-FLOOR PLAN

Sou Fujimoto's new building, like his others on the same premise – the 7/2 House, where two apartments are joined under seven pitched roofs (hence its name), and a dormitory in a sinuous, wiggling shape like that of a caterpillar – is a head-turner. The contrast with the more conventional institutional buildings on the same site, a 14,590-square-metre (157,000-sq.-ft) plot overlooking the city of Date, about 80 kilometres (50 miles) south of Sapporo, could not be stronger.

As one of the winners of the *Architectural Review*'s Awards for Emerging Architecture in 2006, this building brought its young architect into the international limelight. Following 'an infinite, strict and artificial design process', the facilities for psychologically disabled children and their staff – including the children's dormitory, kitchen and canteen, library, multi-purpose rooms, staff rooms, and consultation and waiting rooms – are housed in an apparently random cluster of white cuboids.

For Fujimoto, the current architectural trend for transparency and smooth curves is problematic, as it tends to stop its users from registering their surroundings. Obstacles, on the other hand, heighten awareness. In this centre, Fujimoto creates a space that opens gaps, forming alcoves and corners with no particular function. Out of this, he has created an environment that is not only as cosy and comfortable as a family home but also as exciting and unpredictable as a big city, such as Tokyo – where Fujimoto, who grew up in a town similar to Date in midland Hokkaido, now lives and works.

The gaps created by spreading out two-storey cuboids of similar dimensions have been filled with one long, meandering double-height walkway or 'street', to function as communal space. The interstitial space narrows and expands, at times unfurling into courtyard-like spaces brimming with natural light. Two entrances – one solely for the children's use – mark the beginning and end of this walkway, and timber-framed stairways are attached to the walls. By partially exposing the air-conditioning ducts behind the ceiling in the communal space, Fujimoto also manages to suggest continuity rather than an end to the available space beyond.

The building's structure, despite the use of such a rigid material as reinforced concrete, thus leaves some ambiguities through disjuncture. The children can, for example, peer into a room tucked inside an opening that resembles a fireplace in a giant chimney, although they probably find nothing there but their own imagination. Fujimoto's innovative structure encourages the children to have a less formal but more inquisitive spatial experience.

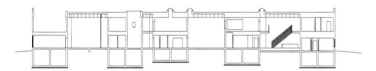

SECTION

OPPOSITE, LEFT The centre is a cluster of white cubes. OPPOSITE, RIGHT Fujimoto's other work on the site – 7/2 House – is visible in the background. BELOW, LEFT TO RIGHT A double-height communal space; one entrance is reserved for the children; gaps are fitted with clear glazing. BOTTOM LEFT An upper-level landing. BOTTOM RIGHT The gymnasium, also designed by Fujimoto, can be seen through the window on the left.

HEALTH AND RELIGION 177

Leaf Chapel

Kobuchizawa, Yamanashi Prefecture

KLEIN DYTHAM ARCHITECTURE, 2003–2004

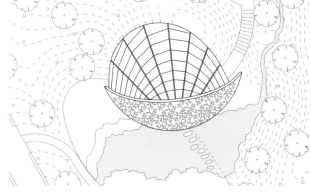

SITE PLAN

A non-denominational sanctuary for weddings, Leaf Chapel sits in the grounds of a 1980s vintage resort modelled on an Italian hill town but located at the edge of the snow-capped Japanese Alps. One of some forty hotels owned by the 'golden boy' of the Japanese hospitality industry, the compound – besides needing a revamp – was missing an attractive venue for the exchanging of vows. So the hotelier commissioned Astrid Klein and Mark Dytham, whose Tokyo-based firm is known for inventive and eye-popping design, to create a suitably photogenic chapel. The success of their building – which, on propitious days, hosts as many as ten weddings – was the beginning of a blissful partnership, and the project was the first of numerous improvements the architects have made to the hotel, including the Brillare Dining and Party Room and the Moku Moku Yu bathhouse (pages 96–99 and 112–13).

Taking their cue from the client's love of outdoor weddings, the architects placed their 176-square-metre (1900-sq.-ft) building in the hotel's manicured garden. The concept for the chapel grew out of the image of a pergola: it is an intimate one-room space sheltered by two 'leaves', one of opaque glass and one of perforated steel, that shield the ceremony from view but also dramatically open to the surrounding landscape.

A stepped exterior path leads from the hotel to the chapel. Guests enter the sanctuary, where they are seated on curved pews of dark wood and clear acrylic with a floral imprint. Meanwhile, the bride and groom slip in through a side door leading to a sequestered waiting area, and it is from here that they make their grand entrance. Dressed in white, the bride stands out dramatically against the blackened pine rear wall and black granite floor as she glides to the front of the room, where the ceremony takes place against the backdrop of the steel leaf. Lined with fabric, the curved enclosure is dotted with a stylized vine pattern composed of 4700 perforations, which bathe the room in a soft, milky light and transform the metal wrapping into a billowy swathe of lace.

At the end of the ceremony, as the groom kisses his new wife, the steel cover lifts magically like the bridal veil, opening the room to the garden in just thirty-eight seconds and symbolically launching the happily married couple into their new life together.

RIGHT, TOP, AND OPPOSITE, TOP Sequential views of the roof opening. RIGHT, BOTTOM Roof detail and pews. OPPOSITE, BOTTOM LEFT Interior view, out towards the pond. OPPOSITE, BOTTOM RIGHT The chapel at dusk.

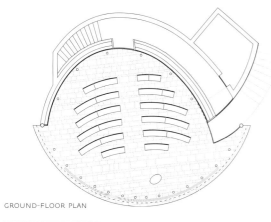
GROUND-FLOOR PLAN

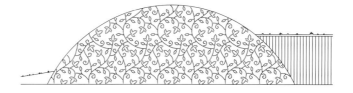
ELEVATION

HEALTH AND RELIGION 179

Sendai Baptist Church

Sendai, Miyagi Prefecture

SOY SOURCE ARCHITECTS, 2006–2007

ELEVATIONS

 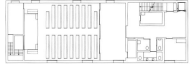

GROUND- AND FIRST-FLOOR PLANS

BELOW Apartment towers and mixed-use buildings surround the modest church. OPPOSITE, TOP LEFT The masonry was coloured with red dye, and a styrofoam layer imparts a rough texture. OPPOSITE, TOP RIGHT View from the playground. OPPOSITE, BOTTOM The sanctuary.

Like people, buildings subjected to the ravages of sun and snow do not always age well, and in Japan, frequent earthquakes add an extra dimension. When the fifty-year-old Sendai Baptist Church began to show signs of wear and tear, its congregants knew they could not simply let nature run its course. While some members hoped a little reconstructive surgery would do the trick, others envisioned replicating their beloved building. Because the old church's structural system was lacking and it no longer fitted the needs of the 120-strong congregation, all eventually agreed to commission the local firm Soy Source and start from scratch.

At the time of the original church's conception, Sendai was a quiet agricultural town dotted with low-rise buildings, but today the site sits in a dense mixed-use neighbourhood, hemmed in on three sides by nondescript high-rise flats. The architects chose a simple but bold rectangular massing to stand out against this backdrop, and decided to build with concrete. It was not to be just any concrete, however: to create a strong impact, the masonry was coloured with red dye to mask its grey pallor. In addition, its formwork was padded with a styrofoam layer that imparted to the finished surfaces a rough texture capable of concealing age spots and unsightly lines, should they ever appear. However, the architects expect the building to mellow and improve as would a traditional wooden temple or shrine.

A solid wall of warm, welcoming reddish-brown concrete forms the façade. Devoid of doors and windows, it is adorned only with a simple steel cross. The building is set back from the street with its entrance at the side, opening on to a large playground along the west face. A combined kindergarten and community room fills the ground floor. Stairs lead up to the double-height chapel with an altar at the far end and a baptistery pool beyond, and a modest apartment for the resident clergy occupies the top floor. In lieu of windows, two roof gardens fill the living quarters with daylight.

Sheltered by concrete walls, the two gardens pierce the church's pitched roof. Intended to minimize the accumulation of snow and maximize sunshine to the playground, the roof ridgeline follows the building's diagonal axis. Inside, this creates an awe-inspiring place of worship with a ceiling soaring to 6 metres (20 feet) at its highest point, evoking venerated church architecture.

While its thick walls protect the spiritually charged sanctuary, whimsical square windows puncture the church's thick skin and tie it to the here and now.

By recycling original furnishings, such as the pews and altar, and reusing roof tiles as floor pavers, Soy Source anchors its building to the past. But at the same time, its concrete enclosure positions the church well for the future.

HEALTH AND RELIGION

White Chapel
Osaka City, Osaka Prefecture
JUN AOKI & ASSOCIATES, 2005–2006

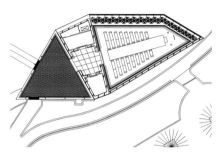

SITE PLAN

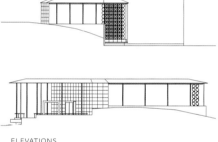

ELEVATIONS

Most people in Japan practise a mixture of Buddhism and Shintoism, yet many young couples have their hearts set on a church wedding with all the trimmings. Since the only legally binding marriage ceremonies in Japan take place in municipal government offices, wedding chapels are just for show, but when it comes to pomp and celebration they are the answer to many a young couple's prayers. Catering to this lucrative market, many major hotels countrywide maintain built-in chapels, which book up years in advance. When the Hyatt Regency Osaka was remodelled in 2006, management reasoned that one chapel was good but two would be better, and engaged the Tokyo architect Jun Aoki to build a second.

Having designed numerous boutiques for the luxury brand Louis Vuitton, Aoki brought an eye for glamour and fashion to the project. Impeccably dressed with a filigreed wall of interlocking rings, his White Chapel is a free-standing pavilion perched beside the hotel's carp-filled pool. Although the symbolism of the rings is hard to miss, the delicate assemblage belies its structural might. Made of pure steel, the rings work together to support the chapel's shimmering aluminium roof.

The jewel of the hotel complex, the chapel is contained in an elegant, faceted volume. Accessed from the main lobby by footpath or bridge, it is composed of a covered triangular terrace, a wedge-shaped entry foyer and a tapered sanctuary, which accommodates seventy guests. The three pieces are strung together axially: the centreline of the terrace connects to that of the foyer, which in turn connects to that of the sanctuary, where the aisle, flanked by Aoki-designed pews, culminates at the altar.

Equally rigorous geometry guided the design of the ethereal outer wall. Its crystalline framework is made of 1500 steel loops arranged in stackable tetrahedrons. Aligned in orderly rows, the three-dimensional units perform well under pressure while contributing to the building's porosity and attractive appearance. The metal scaffolding is embraced by sheets of glass on one side and by diaphanous organdie fabric, custom-made by textile designer Yuko Ando of Nuno Corporation, on the other. While the clear glass reveals the intricate metalwork's decorative character, the gauzy white cloth veils it, lest Aoki's architecture upstage the bride and her retinue.

ABOVE The approach to the chapel is over a carp-filled pool. OPPOSITE, TOP LEFT The sanctuary is enclosed by an ethereal wall of steel rings. OPPOSITE, TOP RIGHT Detail of the exterior wall. OPPOSITE, BOTTOM LEFT The sanctuary. OPPOSITE, BOTTOM RIGHT The chapel at dusk.

HEALTH AND RELIGION 183

186 Big Window House
188 Engawa House
190 G
192 Gae House
194 Hall House 1
196 House Crane
200 House N
202 House O
204 Kamakura House
208 Lotus House
210 Moriyama House
212 N
214 Natural Strips II
216 Rectangle of Light
218 Ring House
220 Rooftecture S
222 Stage House
224 Steel House
226 Tetsuka House
228 Y House
230 Yakisugi House

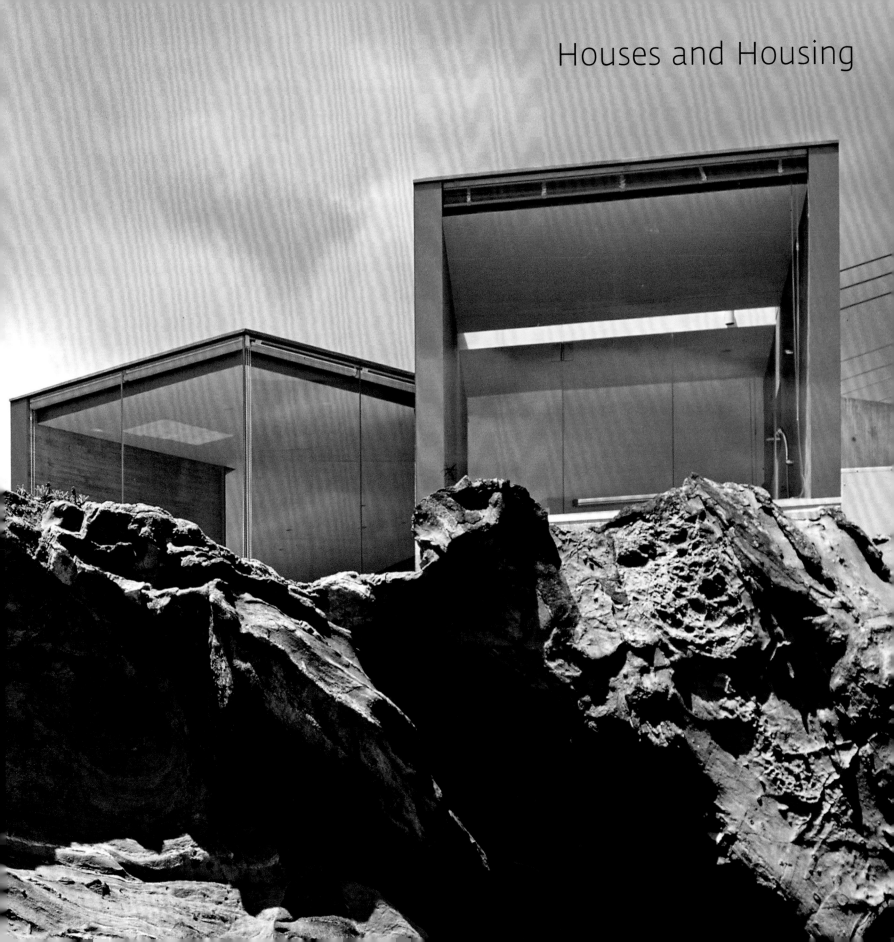

Houses and Housing

Big Window House

Yokohama, Kanagawa Prefecture

TEZUKA ARCHITECTS, 2003–2004

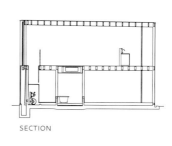
SECTION

GROUND-FLOOR PLAN

FIRST-FLOOR PLAN

This is an extremely simple building, but one executed with such aplomb that it becomes a demonstration of architectural depth and sophistication. A two-storey structure opposite a park in a dense urban setting, offering 95 square metres (1020 sq. ft) of floor space and designed over a period of five months, it is called Big Window House with some justification: an electrical mechanism causes half of the principal façade to disappear, opening the entire upper storey to the world outside.

The house is accessed from a side door. Essentially a 'one-up-one-down', and rectangular in plan, it has bedroom and bathroom facilities on the ground floor, and kitchen, dining and lounge spaces on the first. The key to the design is the two-storey slot at the front, accommodating the staircase and an enclosure into which the large window drops. Light filters easily through to the ground floor, while the elevated glazing draws visitors upwards to where the view to the south is framed by the building itself. The house is certainly an exercise in compact living: storage is provided by full-height units at the rear and cupboards under the stairs; many of the doors slide, rather than swing, open; and the function of each space is virtually preordained.

This flat-roofed ensemble is one of a number of challenging homes to emerge from this husband-and-wife team of architects. Other examples of their work in this sector include the Wall-less House (a three-storey structure supported on just a central core and two very thin columns); the Canopy House, which features a roofline so wide that large windows can remain open even during heavy rain; and the Roof House, a single-storey building with an open-air facility for dining and lounging located on the 'flat' (but actually slightly sloping) roof.

Tezuka Architects appear to be wrestling with the house as a social, as well as architectural, construct. By asking such fundamental questions as 'What is a window, or a roof, or a wall?', they continue to open up apparently endless possibilities for living.

OPPOSITE Big Window House occupies a small site in a dense urban setting. BELOW, LEFT AND CENTRE, AND BOTTOM LEFT AND CENTRE The glazing that covers the upper half of the principal façade can disappear completely. BELOW, RIGHT A slot contains a staircase parallel with the façade. BOTTOM RIGHT Entrance hall, with under-stair cupboards.

HOUSES AND HOUSING 187

Engawa House

Adachi, Tokyo

TEZUKA ARCHITECTS AND MIAS,
2002–2003

Unlike the more hip areas of Tokyo, such as Setagaya, the north-east suburbs of Adachi are still plagued by the dominance of 'housemaker' homes, which stand with assembly-line regularity. Architects rarely venture out to this part of the city, but it was for that very reason, the husband-and-wife team Tezuka Architects proudly maintain, that they took on this job.

The brief was interesting, too, for it captured the prevalent Japanese phenomenon: to create a new arrangement for *nisetai jutaku*, 'a house shared by two generations'. The Tezukas themselves live in such a house, so they understood very well the need for privacy as well as proximity between the two sets of relations. The old arrangement, in which the young client with his ageing mother-in-law and his own growing family lived under one roof, was beginning to feel cramped, so when the adjacent plot became available for purchase, he did not hesitate for a moment.

The new house is a simple rectangular box, 16.2 metres long, 4.6 metres wide and 3.5 metres high (53 by 15 by 11½ ft). The exterior is clad in vertical strips of timber, while the interior space is broken up by screens, which do not reach the full height of the building; kitchen odours and steam from the bathroom rise into the roof void and exit via the high clerestory windows that line the street side of the building. It is a house of two personalities: from the street, where the windows are set 2.2 metres (7¼ ft) from the ground, it presents a distinctly private face; but the other façade suggests openness and communal living.

The garden became the focal point for this project, for it is here that the two families converge. The relatively low structure of the house does not shade the garden, which has become an essential third space through which, for example, the children hop over to their grandmother's house in the evening to watch television. The house's name refers to the porches on the garden side of traditional Japanese houses; in this updated version, nine large sliding doors allow one side of Engawa House to open almost completely to the garden (the large steel beam that enables them to slide effortlessly is hidden from view).

The house was designed around ideas of traditional European domestic life: log fires, family meals, views into gardens. But it also has a strong Japanese flavour, particularly in the way that it remains a single space in spite of its subdivisions. As one half of the home of an extended family, it operates very well. Tezuka Architects have perfectly blended old and new, with the garden acting as an unwitting intermediary to absorb any generational tension that may have existed before.

OPPOSITE, TOP The subtle street elevation. OPPOSITE, BOTTOM Engawa House at dusk. THIS PAGE One wall slides away, opening the house entirely to the garden – the focal point of the development.

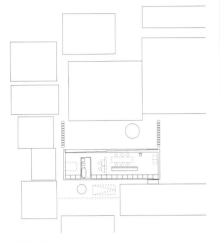

SITE PLAN

FLOOR PLAN

SECTION

G

Meguro, Tokyo

JUN AOKI & ASSOCIATES, 2003–2004

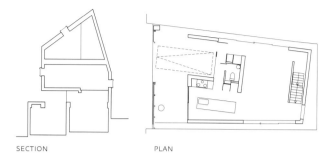

SECTION　　　　PLAN

A private home for a graphic designer and his family, G melds comfortably with its surroundings, a typical residential neighbourhood in central Tokyo. But somehow, at the same time, it does not fit in at all. It has many trappings of the conventional house – a pitched roof, off-the-rack operable windows, and even patterned wallpaper in the living room – but, true to form, Aoki cleverly uses these elements to create a fresh, light-filled interior and an exterior that pokes fun at its neighbours in a friendly way.

As is common all over Tokyo, plots in this area are small and the houses fairly close together. Because of the tight conditions, wall openings have to be carefully choreographed lest the goings-on in one house become perceptible in another. In this setting, windows alone rarely provide enough daylight, and skylights are the sensible conclusion for many homeowners. For Aoki, however, they were the logical beginning.

G is in two distinct parts: a reinforced-concrete plinth topped with a wooden frame structure that looks like a house but acts as a giant skylight. A narrow gap separates them, but links interior to exterior where glass inserts offer a glimpse of trees. The top section is set back, while the bottom is only 1 metre (3 ft) from the next house – a polite nod to the time when both properties were owned by the same person. But apart from their proximity and matching paint colour, there is no connection between the two homes.

Unlike the house next door, G abuts the street with a white forecourt where the clients stow their car (since vehicle owners in Tokyo must possess off-street parking). The forecourt leads into the base section of the building, which contains the kitchen and the living/dining room. A soaring space defined by wooden columns and beams above, the living area links the top and bottom sections volumetrically, but stairs connect physically to the private quarters: a children's bedroom and family study area one floor up and the master suite above that.

In functional terms, the upper portion, which gives a feeling of top-heaviness to the structure, is bigger than necessary, since it holds only bedrooms. But it bathes the entire interior with soft daylight, which filters down through the sixteen operable skylights dotting the walls and roof. Although the square shape and modest size of the openings mimic the windows of houses near by, their seemingly random placement offers no clues about the house's inner workings. Instead of correlating them with rooms, Aoki located them to provide illumination and ventilation and to maintain privacy. Practical considerations also determined the shape of the roof; although house-like, it was influenced by the city's stringent shadow-limiting laws.

With G, Aoki disassociates his forms from their ingrained iconography, but he does not discard such markers altogether. The result is a contextual response that gently challenges the status quo.

OPPOSITE The house abuts the street with a white forecourt. RIGHT Detail of the wooden frame. FAR RIGHT AND BELOW, RIGHT Stairs lead up from the communal areas to the private quarters. BELOW, LEFT Daylight filters down to illuminate the living spaces at the base of the house.

HOUSES AND HOUSING 191

Gae House

Setagaya, Tokyo

ATELIER BOW-WOW AND TSUKAMOTO LABORATORY,
TOKYO INSTITUTE OF TECHNOLOGY, 2001–2003

Like many of the small dwellings featured in Atelier Bow-Wow's *Pet Architecture Guide Book* (2002) – which was also produced in collaboration with graduate students from the Tsukamoto Laboratory – Gae House is built on a tiny site. The miniscule plot, not much more than 70 square metres (750 sq. ft) but conveniently just a stone's throw from such fashionable areas as Shibuya and Jiyugaoka, was left over from the division of a larger site as a result of a new inheritance ruling.

The architects' challenge was to house an impressive collection of books, which the client, a well-known literary critic, calculated to require about 50 metres (165 ft) of shelving. The problem was further complicated by the strict building-to-site ratio enforced in this highly congested area of Tokyo. Given the tiny size of the plot, the house would have to be very small indeed.

When it comes to maximizing the potential of a tight urban space, however, Atelier Bow-Wow is at the top of its profession. For this house, the architects eventually arrived at a solution that may seem strange to Western eyes; however, the top-heavy structure of Gae House resonates powerfully with the Japanese temple vernacular, in which the space under the eaves displays the impressive criss-crossing beams that support the roof.

The enormous roof and protruding eaves of this building, which has been dubbed the 'mushroom house', nevertheless serve practical purposes. Horizontal glass panels tucked under the eaves allow light into the top floor, where the kitchen, dining and living areas are situated, and the client's beloved Volkswagen Golf is protected from bad weather under the shelter they provide.

The basement houses the client's books, but in order to bring light and fresh air in while retaining a generous amount of wall space for the shelves, a large hole was carved out of the ceiling. Despite depriving the owner of about half the available space on the ground floor, it meant that he could concentrate on his work in the den-like study. The galvanized-steel ceiling in the top room disperses light, resulting in an iridescence in which the upper floors are highlighted as the social spaces, for eating, washing and entertaining; the basement, where light is absorbed and views restricted, encourages meditation, daydreaming and recuperation.

ELEVATIONS

SECTIONS

OPPOSITE Gae House's extended eaves protect the owner's cherished car from bad weather. RIGHT, TOP AND BOTTOM A galvanized-steel ceiling disperses light on the top floor. FAR RIGHT, TOP AND BOTTOM A large hole in the ground floor solves the problem of storing the client's large collection of books and brings light and air into the windowless basement.

Hall House 1
Otsu, Shiga Prefecture

ALPHAVILLE, 2005–2006

In Japan, where the secondary market for houses is practically non-existent, personal preferences – no matter how eccentric – often guide residential design. Commissioned by a couple who would rather hang out than go out, Hall House 1 was built around a billiard table. Requiring a large, column-free, sound-proof space, this request was far from insignificant, especially on the clients' tight budget and oddly shaped site.

Located in a suburban community but on leftover land hemmed in by residential developments, the house greets the street with a poker-faced concrete wall that reveals nothing of the interior. Even the front door is concealed within the covered car park appended to the side. Inside, the house shows its hand. Essentially it is a one-room dwelling with a pentagonal plan that maximizes the peculiar geometry of the site. Hall House 1 contains a remarkable range of functions and a rich variety of spaces, all connected by a spiralling circulation path.

The foyer, with its full-height partition, leads into the home entertainment centre, a shadowy place reminiscent of a pool hall. The enormous billiard table stands in the middle, where there is plenty of room to manoeuvre the perfect shot. The back of the foyer wall doubles as a screen for watching the clients' extensive film collection, which is housed in wall-mounted shelves. Opposite the screen, broad stairs lead to the open, steel-clad kitchen half a level up, but also serve as theatre seating. Narrow steps continue the circuit by leading to the loft-like sleeping area and the bathroom. An exterior ramp continues up to the roof.

Abutting an outdoor terrace with a wall of clear glass, the bathroom is the primary source of daylight for the entire house. Although it is orientated towards the school playground across the street, a translucent glass wall surrounding the plant-filled patio shields it from view. The only other sources of natural light are the front door and a frosted-glass window in the kitchen. As a result, the interior is fairly dark, but this was a reasonable trade-off for privacy.

The internal functions as well the external constraints are reflected in the building's multi-faceted, entirely concrete enclosure, its angled walls complying with the property lines. The hipped roof accommodates the sleeping quarters; composed of three folded planes, it teams up with a single mighty beam, disguised by the bed headboard, to carry the structural load.

Despite being dealt extremely difficult cards, the architect – a young firm based in Kyoto – has played the game well, turning a problematic site into a convenient, personalized retreat for two.

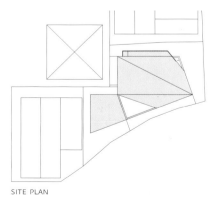

SITE PLAN

OPPOSITE, TOP The house greets the street with a blank concrete wall. OPPOSITE, BOTTOM Roof detail. Privacy on the accessible rooftop is ensured by translucent glass panels. RIGHT, TOP The glass-enclosed bathroom acts as a giant window that supplies daylight to the whole house. RIGHT, BOTTOM The kitchen overlooks the living room, which is dominated by a pool table.

SECTION

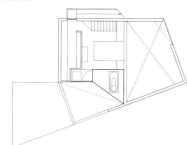

FIRST-FLOOR PLAN

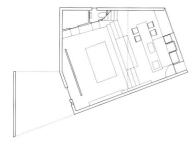

GROUND-FLOOR PLAN

HOUSES AND HOUSING 195

House Crane

Karuizawa, Nagano Prefecture

ATELIER BOW-WOW, 2005–2007

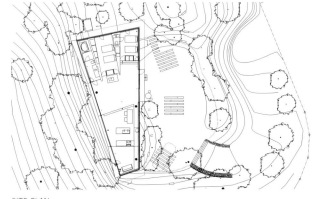

SITE PLAN

All the trappings of a classic Case Study House – floor, roof, columns and wraparound glass – are here assembled with a twist by Atelier Bow-Wow. While the stepped floor plane follows the lie of the land, the house's columns resemble randomly placed tree trunks and its roof tilts at different angles all along its length. Instead of imposing an artificial order on its environment, House Crane takes its cue from nature.

The site, on top of a small hill hidden in the woods, in a planned second-home development in this exclusive resort town, is sequestered both from the neighbours and from the crowds who flood the town all summer long, and belongs to a couple from Tokyo who wanted a holiday house open to the landscape in every direction. Although their property has no dominant view, the house is perched on the edge of a precipice, overlooking the canopy of trees rooted in the steep slope on one side but easily approached from the relatively flat land on the other.

Encased in splayed sheets of floor-to-ceiling glass that point towards the decline, the wedge-shaped house is entered from the side. It contains a series of functional areas differentiated primarily by changes of level, descending gradually from the wider end. Sliced lengthwise to avoid corridors, this end holds the bedroom, guest room and bathroom separated by partitions, each on a slightly different level but all facing the expansive living room with its wood-burning fireplace. From there, more steps lead down first to the dining and kitchen area and then to the den, a cosy seating area at the tip of the house.

Although the incremental changes between levels are small, they add up to an overall drop of more than 1 metre (3 ft) end-to-end. Unsurprisingly, this difference had strong implications for the design of the roof. The easy answer was to slant it downwards, but the appearance that would give was too strong for the site; also, the architects wanted to frame the views with horizontal eaves. Instead, they kept the roof level, varying the ceiling heights and creating a unified but complex, composite plane.

It unquestionably took much effort on the part of the architects to mould House Crane to its environment. But the result, despite being termed by them a 'weak' building, is more powerful than most 'strong' ones.

ELEVATIONS

OPPOSITE, LEFT Encased in glass, the house opens to the landscape in every direction. OPPOSITE, RIGHT The narrow tip of the wedge-shaped home perches on the edge of a precipice. BELOW, LEFT Irregularly placed columns echo the tree trunks outside. BELOW, RIGHT Broad eaves shade the house from the afternoon sun. BOTTOM LEFT The bedroom is at the back of the house. BOTTOM RIGHT The stepped floor plane follows the site's natural topography. OVERLEAF Cabinetry and changes in floor level define the kitchen and dining areas in the middle of the house.

HOUSES AND HOUSING 197

House N

Oita City, Oita Prefecture

SOU FUJIMOTO ARCHITECTS, 2006–2008

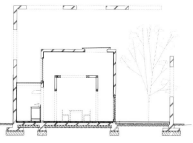

SECTION

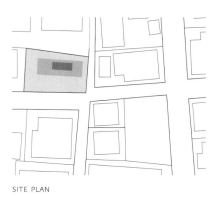

SITE PLAN

Located on Japan's southern island of Kyushu, House N is the product of Sou Fujimoto, a Tokyo-based architect who grew up on the northern island of Hokkaido. While winters in Fujimoto's home town are harsh, warranting thick walls, the intense sun and mild temperatures in the south invite a relaxation of the boundaries between inside and outside. A collection of three 'nesting boxes', each with a semi-permeable skin, House N takes a fresh approach to the Kyushu climate. Its carefully orchestrated wall openings greet the street warmly while shielding a gloriously airy and uplifting interior from view.

The project began when a retired couple asked Fujimoto to renovate their home in central Oita, a city of 470,000 people in the middle of the island. But they changed course midstream, and decided to rebuild altogether. Fujimoto responded by covering most of their 237-square-metre (2550-sq.-ft) site with a vast but transparent shell. It encloses a pebble-studded yard with parking for two cars on one side and a wooden deck on the other. In the centre Fujimoto placed a second, smaller shell. Drawing a thin line between interior and exterior, he filled its openings with thick glass sheets that keep out wind and rain but let in light and views. That shell holds the foyer, a *tatami*-floored guest area and the couple's sleeping quarters. The heart of the house is the combined living/dining room, which occupies the smallest and most intimately scaled shell. It abuts the kitchen; separate bath and toilet areas fill a narrow band of space at the back of the house.

While the placement of one volume inside another has created corridors and functional areas, the openings in their walls link the whole into a single, loft-like space. The number of voids varies, but enclosure and exposure are well balanced in every roof and wall. Fujimoto used geometry to fix the measurements and proportions of the openings – there are three different sizes per shell, each a golden rectangle – but he let sunshine and sight lines guide their placement. Some openings filter the sun's rays while framing picture-perfect views of the deep-blue sky; others overlap to reveal glimpses of the street as they conceal the interior from passers-by.

Abstract and devoid of human-scaled elements, Fujimoto's stark white boxes are slightly out of step with the conventional pitched-roofed homes lining the neighbourhood's narrow streets and pedestrian passageways. But their nesting configuration straddles urban and residential scales, tying House N securely into the concentric organization of the city.

ABOVE, LEFT Composed of starkly white, abstract boxes, House N stands out from its suburban neighbours. ABOVE, RIGHT Carefully orchestrated wall openings greet the street warmly but preserve privacy. OPPOSITE, TOP LEFT The house consists of three boxes, which nest one inside the other. OPPOSITE, TOP RIGHT Openings in the exterior ceiling cleverly reduce its voluminous measurement, which would otherwise not fall within the set building coverage ratio. OPPOSITE, BOTTOM LEFT A wooden deck sits inside the biggest box; mature trees grow skywards. OPPOSITE, BOTTOM CENTRE A wraparound window in the dining area is unglazed. OPPOSITE, BOTTOM RIGHT Views out of the master bedroom are fragmented.

House O

Minami Boso, Chiba Prefecture

SOU FUJIMOTO ARCHITECTS, 2006–2007

House O perches behind another house on the edge of a jagged inlet on the south coast of Boso Peninsula, completely invisible from the street. Despite its proximity to Tokyo, this small seaside town feels as though it exists in another time. *Ama* (traditional female divers) still dive for clams and shells here, wearing only *shiroshozoku*, their traditional white cotton diving outfit, and a pair of goggles; and a flagpole next to House O alerts them when the waves are too high for safety. The owners of the house – a doctor and his wife, who come to this area every year to fish – found the site by chance.

In such a setting, this low-rise house clad in exposed concrete appears uncompromising, even brutally cut off from its surroundings, but the architect tamed not only the crashing waves of the Pacific Ocean but also the rock-solid mindset of the clients, who were expecting a Miesian rectilinear glass box that would have taken in the ocean view in one sweep. Instead, the final model, sliced up and joined back together at various angles, in its wild unpredictability poetically echoes the surrounding craggy rock formations.

The result is a tightly proportioned continuous space that branches out in eight different directions, weaving in and out of crevices and corners, each branch fulfilling a different function. The kitchen units are hidden behind a counter at one end of the house's 'spine', which then pans across the dining and living rooms, zigzagging along the glass wall panels, which are joined seamlessly to obtain the full visual effect of the ocean just a few metres away.

With no eaves to protect the house from sun or rain, the feeling of being exposed to the elements is intensified. The harshness of the sea is tangible, even inside the house. The seamlessness is further emphasized by the horizontal patterns on the interior walls, created by casting concrete against Japanese cedar boards 5 centimetres (2 in.) thick. The timber doors were painted to emulate the casting and complete the effect.

The snaky contour of the spine terminates in the master bedroom. Along the way, a concrete box encasing a sunken bath juts out, framing another magnificent sea view perfectly. The heavy concrete rendering could have suppressed such a sensitive evocation of the house's interconnectedness with the outside world; but instead, the couple can enjoy the special meandering promenade, which presents new and unexpected vistas of the dramatic panorama, all year round, in sun or rain.

ABOVE House O seen through the craggy rock formations. OPPOSITE, TOP Aerial views reveal the remarkable structure of the house. OPPOSITE, CENTRE LEFT A skylight brings a beam of light into the master bedroom. OPPOSITE, BOTTOM LEFT The main entrance. OPPOSITE, BOTTOM RIGHT Very little comes between the study and the Pacific Ocean.

SITE PLAN

HOUSES AND HOUSING 203

Kamakura House

Kamakura,
Kanagawa Prefecture

FOSTER + PARTNERS, 2000–2004

Protected by a row of security cameras and accessed only through a state-of-the-art fingerprint-recognition system, Kamakura House stands amid outstanding natural beauty in the hills near the coastal town of Kamakura, south of Tokyo. Apart from one skyscraper (Century Tower), this is the second house that Foster + Partners have designed for the late Kazuo Akao, former owner of the Obunsha Group, a well-known publisher.

The London-based architect is not known for designing private houses, but its hallmarks of attention to detail and clarity of planning are just as appropriate here as they are in airports and large offices. A highly polished rectilinear composition of parallel spaces sits on a sloping plot enclosed by rugged, tree-covered, cave-dotted hillsides. Samurai swords were once manufactured on this auspicious site.

The house incorporates a Shinto shrine, perhaps in honour of such a dramatic past, while a separate pavilion provides ample space for holding business meetings and for displaying the publishing magnate's impressive collection of Buddhist art. Both buildings have been crafted to enhance the sense of harmony that characterizes the site, framing a landscape that is rich in historic and cultural associations.

Like many domestic buildings in Japan, Kamakura House is essentially inward-looking, especially since the cliff walls give the site a courtyard-like appearance. Openings capture key views, focusing on a mature cherry tree, and light and shadow are carefully controlled, as are the ways in which Akao's collection of contemporary artworks – including the work of Gilbert & George – is presented, by slicing up the rectilinear box and staggering its parts.

Kamakura House is more than just a backdrop of pared-down modernity, however: materials and texture come into play to create a sense of luxury. The design team developed a number of bespoke materials for the project, including custom-made reconstructed stone for the primary walls and glass blocks made from recycled television tubes. The interior also features hand-sculpted terrazzo, antique Chinese floor tiles and glazed volcanic stone for the indoor pool.

A lighting system, including fibre-optics and dedicated spotlights, backlights the glass blocks and reinforces the architectural containment. The house is a tour de force and a work of art, a foil for nature but also – like a samurai sword – cutting through it.

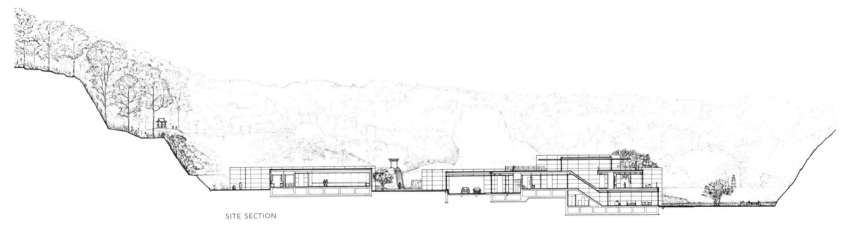

SITE SECTION

SITE PLAN

OPPOSITE, TOP LEFT The coastal town of Kamakura is seen in the distance. OPPOSITE, TOP RIGHT Kamakura House is a finessed Modernist building. OPPOSITE, CENTRE RIGHT Parallel architectural spaces nestle on the hillside. OPPOSITE, BOTTOM RIGHT Cliff walls make the site into a natural courtyard. TOP LEFT Materials and texture create a sense of luxury. TOP RIGHT The house encloses the traditional meditation room. BOTTOM LEFT Interior and exterior spaces merge. BOTTOM RIGHT A statue of Bishamonten (God of War and Warriors) stares back from another pavilion. OVERLEAF Glass blocks created from recycled television tubes draw in daylight at the end of the corridor, while backlit resin moulds embedded in the stone wall artificially illuminate it from the side.

HOUSES AND HOUSING 205

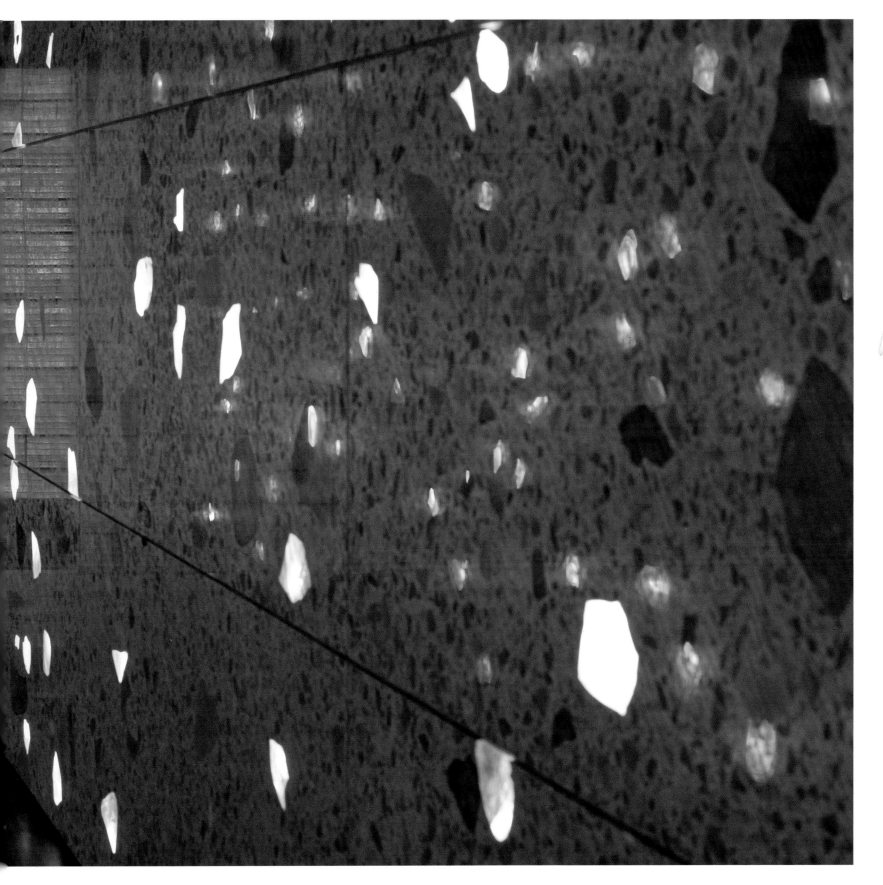

Lotus House
Kanagawa Prefecture

KENGO KUMA & ASSOCIATES, 2003–2005

Positioned within a private estate, Lotus House sits like the elegant palaces of the Heian period, spreading its symmetrical wings in front of a pond peppered with the lotus flowers after which it is named.

In Buddhism, which is particularly significant for the house's owner, the lotus flower conjures up an image of paradise (the enlightened Buddha sits and meditates on one). The client carefully relocated and restored a Muromachi period (1338–1573) Buddhist meditation hall and the traditional *soan* (grass hut) tea house on his estate, so that he could authentically meditate and perform tea ceremonies to calm his mind on his weekends away from the city.

The theme of paradise is also reflected in the façade, which is a chequerboard of blocks of travertine stone so thin that, suspended by flexible stainless-steel bars, they 'flutter like delicate lotus petals'. The water of the pond dissolves the image of the house in a shimmer of light, and the façade also contributes to the collapsing of boundaries between inside and outside.

Reminiscent of traditional Japanese houses, the complex contains multiple routes; the main entrance is distinguished by another pool, which reflects the chequered façade. Stone steps, steel staircases and sliding glass doors all play a part in decreasing the weight of the structure. Kuma further dematerializes the house by employing porous steel mesh inside: it shapes the spiral staircase, for example, then spills over to the ceiling of the kitchen and dining room. A similar effect is achieved in the study, where a narrow steel staircase allows access to the top shelves of the bookcases. The chipboard ceiling counteracts the feeling of luxury, which in the teaching of Buddhism is all part of an illusionary world.

The courtyard, which contains a few pieces of antique furniture only, functions as a transitory space, separating the activities of the everyday – cooking, washing and sleeping (all accommodated in one wing of the house) – from the more refined, leisurely activities – reading, playing the piano and watching films – contained in the opposing wing.

In its ethereal construction, Lotus House offers enclosure without containment. This ambivalence is illustrated by the two baths that sit next to each other, one enclosed in glass overlooking the other, an outdoor one, bounded only by a chequered-stone screen. Just as he can swap the inside bath for the outside one, the owner can slip between luxuriating in the pleasures of life and living in a more abstemious way – a surer way to paradise.

OPPOSITE, TOP The chequered façade and the shimmering pond soften the boundary between inside and outside. OPPOSITE, BOTTOM LEFT An outdoor bath. OPPOSITE, BOTTOM RIGHT The two wings are arranged symmetrically. RIGHT A sweeping spiral staircase leads up from the kitchen. BELOW, LEFT Façade detail. BELOW, RIGHT One side of the house provides space for leisure activities.

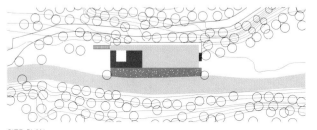

SITE PLAN

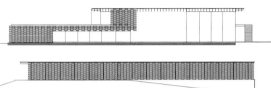

ELEVATIONS

Moriyama House

Ota, Tokyo

OFFICE OF RYUE NISHIZAWA, 2002–2005

Moriyama House, a cluster of ultra-slick white cuboids, is deliberately confrontational. The architect Ryue Nishizawa (a principal of SANAA), who also established his own office in 1997, bemoans the recent loss in Japan of a more relaxed attitude towards privacy and boundaries. Gone are the days when care and commitment went into tending the street outside one's house; instead, all the effort of cleaning goes on within the boundary walls, while the streets are left strewn with rubbish.

Most old boundary walls – such as bamboo *kakine* – which allowed neighbourly chitchat, have been replaced by less porous, often concrete, versions, alienating people from one another decidedly and permanently. The first thing Nishizawa did when he was commissioned to build a series of flats, including accommodation for the client, was to do away with the boundary walls. Furthermore, instead of hoarding all the units under one roof, the architect spread them upwards and outwards, creating, in effect, a small 'village' on the 300-square-metre (3230-sq.-ft) corner site.

Six flats – each with its own unique garden – are formed out of a total of ten units, some of which are pod-like. They all have one characteristic in common: unashamedly large windows (whether sliding, swivelling or opening at one end), which look directly on to the neighbours' houses.

The cleverly laid out accommodation is surprisingly spacious. The client's own home is spread over four separate units and includes a four-storey tower with a bedroom in the cellar, a sitting room on the ground floor, a library on the first floor and a sitting room-cum-gallery at the top. Surrounding the client's own garden are a small pavilion containing a bathroom and another housing a kitchen, which is linked to the main tower by a glass passage – a reminder that every journey made in this 'village' requires the resolution to step out into the world and meet other people. A third pavilion, a double-height living room raised on a podium and facing the garden, also acts as an *engawa* (porch) when its windows are slid open.

A central communal garden above one of the flats is for the use of all tenants. Moriyama House's open and diverse spatial arrangement manages to reintroduce a sense of connection for alienated individuals in one of Tokyo's most densely populated neighbourhoods.

LEFT The free-standing structure of the client's living room is raised 30 centimetres (1 ft) above ground. ABOVE Moriyama House sits on a 300-square-metre (3230-sq.-ft) plot in one of Tokyo's most densely populated neighbourhoods.

TOP LEFT Boundary walls have been removed altogether. TOP CENTRE All the cuboids accommodate large windows that look out directly on to the neighbours. TOP RIGHT The 'gallery room' is at the top of the three-storey tower built for the client. BOTTOM LEFT A glass corridor connects living room and double-height bathroom in one unit. BOTTOM CENTRE The client's kitchen is seen through the walls of another glass corridor. BOTTOM RIGHT The client's free-standing, pod-like bathroom.

N

Kohoku,
Kanagawa Prefecture

JUN AOKI & ASSOCIATES, 2007

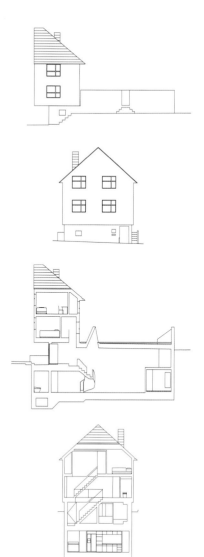

ELEVATIONS AND SECTIONS

Located in one of the many new towns that punctuate the intricate network of commuter railway lines linking Tokyo and Yokohama, N occupies an average plot on a typical street lined with rows of 'housemaker' homes inspired by traditional Western architecture but realized by Japan's unique brand of specialized housing contractors. Composed of two distinct parts – a serious concrete base topped with a caricature of a house complete with mullioned windows and pitched roof – the building coexists amicably with its neighbours but keeps its inner workings to itself.

Instead of the picket fences and ersatz English garden-style trellises belonging to its neighbours, N fronts the street with a solid concrete wall. From the threshold, however, the house reveals its interior at a glance. Steps lead immediately either up or down from the vestibule: straight stairs to the bedrooms in the upper portion, and a sculptural spiral staircase to a single cavernous space that holds a combined kitchen, dining and living area. An expansive wooden deck accessed from an upper landing crowns the partly submerged concrete vault.

Tucked underneath the sloping roof planes, the cosy sleeping quarters are equipped with built-in cupboards and a thick carpet that would be at home in a vintage ranch house. By contrast, the communal space is a 4.4-metre-high (14½ ft) abstract box painted all white. Although it reads as one large room, built-in cabinetry and free-standing partitions designate discrete areas for cooking, eating, studying and entertainment, enabling the family of four to enjoy one another's company while engaged in a range of activities. The only contained space is the Japanese-style room with its wooden walls, *tokonoma* (decorative alcove) and *tatami* floor.

Although windowless and practically underground, the lower level is brightly lit by large skylights and fixtures created specially by the architect in collaboration with the textile designer Yuko Ando of Nuno Corporation. Swathed in gauzy white fabric, giant lanterns hover above the dining and living area. Additional light comes from a cloth-covered, wall-mounted fixture over the kitchen.

From the top of its fake chimney downwards, N is intentionally loaded with quotations from the catalogues of 'housemakers'. But by reordering the grammar that holds them together, Aoki has liberated the house from the standard structure.

OPPOSITE, TOP A pitched roof, a false chimney and mullioned windows parody the typical suburban home. OPPOSITE, BOTTOM View from the street. FAR LEFT The sculptural spiral staircase. LEFT The descending lantern is bespoke. BELOW, LEFT Skylights illuminate the sunken living room. BELOW, RIGHT Occupying a single spacious volume, the communal area is embedded in the earth.

Natural Strips II

Shinjuku, Tokyo

EDH AND MASAHIRO IKEDA, 2003–2005

BELOW At night the house becomes like a lantern. OPPOSITE, TOP, LEFT AND CENTRE Arched, structural steel planes partition the space, and *shoji*-like translucent screens ensure privacy when it is needed. OPPOSITE, TOP RIGHT View towards the kitchen. OPPOSITE, BOTTOM LEFT Steel fins shield the interior and reduce solar heat gain. OPPOSITE, CENTRE RIGHT The stairs are formed from wafer-thin slabs of steel. OPPOSITE, BOTTOM RIGHT The curved steel walls are the primary structural elements.

Commissioned by a retired couple ready for a lifestyle change, Natural Strips II is a private house with a boutique on the ground floor. Set on a hilltop site just five minutes' walk from Tokyo's bustling Shibuya Station (pages 52–53), the small but striking house is clad from head to toe in horizontal metal louvres. This attention-grabbing attire distinguishes it from its immediate neighbours, a mixture of small homes, and from the commercial chaos a little further off. Yet the building's bold appearance was not conceived out of superficial vanity: on the contrary, it resulted 'naturally' when the designers, Masaki Endoh, of Endoh Design House, and Masahiro Ikeda, matched ideal concepts with real conditions.

The clients wanted a glass box open to its surroundings, unlike their previous – concrete – home, but such a thing was practically impossible since the new site was hemmed in by buildings on three sides. As well as privacy, climate control was a concern: glass is notorious for allowing too much heat to build up in summer and escape in winter.

In response, the designers clothed the clear skin with steel fins outside and translucent sliding screens inside. Wrapping three sides of the exterior (the fourth is lined with floor-to-ceiling storage), eighteen louvres hold the glass bands in place and anchor them to mullions. Shaped like the peak of a baseball cap, each louvre does its bit to block out unwanted sun. Inside, milky white, moveable insulation panels clad in sheets of recycled plastic shield the living area from view and create a heat-retaining air pocket.

Naturally, such a thin-skinned house required good bone structure. At first, the architects planned to secure it with a ring of slender columns, which – to allow easy dismantling should the street ever be widened – had to be of either steel or wood. By the end of the design phase they had reduced the support system to just three vertical steel members: two hefty pillars hidden at the back and a single curvaceous 'column' composed of welded plates, standing at the front for all to see.

Functional as well as fine looking, the arched panels also modulate each one-room floor. At ground level they enclose the shop's dressing room; on the first floor, which holds the kitchen and the combined living/dining room, they embrace a Buddhist altar; and on the second floor, which contains the bedrooms and bathrooms, the designers clad the panels' convex surfaces with white tiles and tucked the bath in between.

Verging on the exhibitionist, Natural Strips II feels spacious because of its expansive views and light-filled interior. But it is the wafer-thinness of its various components that truly turns heads.

ELEVATIONS

SECTIONS

FLOOR PLANS

Rectangle of Light

Sapporo, Hokkaido

JUN IGARASHI ARCHITECTS, 2005–2006

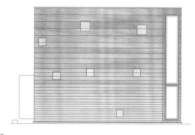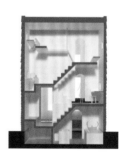

ELEVATIONS AND SECTIONS

Rectangle of Light stands in a sprawling suburb of this northern city next to a long, narrow park with tall trees that protect the area from strong easterly winds. Unusually for Japan, there are very few dividing walls between houses in this suburb, so its outlook is open, inviting and uncluttered: an affable trait when one considers Sapporo's exposure to long and dark winters.

Setting aside the relative spaciousness of the surroundings, Jun Igarashi still had to overcome various limitations, because the young client's tight budget and the strict building-to-site ratio enforced in the area left a mere 40 square metres (430 sq. ft) for him to deal with. The client did not wish to have any large window in the scheme, but Igarashi had to ensure that he did not create a dark, claustrophobic space. The result is surprisingly bright and airy, thanks to a large light well, which illuminates the whole house without compromising the family's privacy.

The exterior gives away very little of what is on offer inside. It is clad in horizontal bands of softwood, the height of which gradually diminishes from top to bottom of the house. The only disruption to this visual harmony is a protruding wooden box, which marks the way into the house. The entrance is painted dark grey, no doubt to accentuate the transition from momentary darkness to the light-filled reception. The spacious main room is half submerged below ground level (the depth of which is indicated by the exposed part of the interior walls), allowing the architect to fit in the tree-house-like children's playroom above it.

The client enthuses about the ecologically sound nature of the house: it remains so bright that he never needs to use artificial light during the day. At the front, small windows are strategically placed to animate the landing, which is used as an office. The back of the house is double-layered, and between the two layers a narrow door and window reaching the full height of the house allow natural light to flood in. A ladder casually propped against the inside wall of the light well offers an alternative means of access to the playroom above, and another ladder continues to the bedroom on the top floor.

The house reflects the quiet, introspective way of life typical of northern Japan. Igarashi, who operates from his home town of Saroma, on the eastern side of Hokkaido, understood well the importance of balancing the need for privacy with the creation of warmth and delight.

OPPOSITE The house is clad in layers of conifer boards; the only opening is at the back. TOP LEFT The kitchen/dining area. TOP RIGHT The light show starts at dawn in the family's bedroom. BOTTOM LEFT The entranceway is painted dark grey, accentuating the transition from dark to light. BOTTOM RIGHT The light well in the kitchen/dining area illuminates the house all day long.

HOUSES AND HOUSING 217

Ring House

Karuizawa, Nagano Prefecture

TNA, 2005–2006

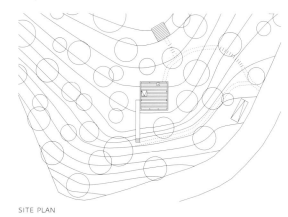

SITE PLAN

BELOW Alternating bands of charred cedar and glass wrap around the house, offering a panoramic view of the surrounding forest. OPPOSITE, LEFT Buoyant stairways rise through the house. OPPOSITE, TOP, CENTRE AND RIGHT The architect-designed kitchen units and iron stove ensure that sight lines are not interrupted. OPPOSITE, BOTTOM CENTRE Wall details. OPPOSITE, BOTTOM RIGHT The master bedroom is on the top floor, for greater privacy.

After encountering problems selling an awkward plot, a young developer who owned a large area of land on top of a hill in the exclusive resort town of Karuizawa decided to call on Makoto Takei and Chie Nabeshima of TNA for a radical solution. The plot, on a steep hill, is eclipsed largely by a road on one side and the view of the back of a neighbouring house on the other. It also lacks a clear view of Mount Asama, sight of which is regarded as an extremely desirable trait for high-end summer homes around Karuizawa, which is a mere hour from Tokyo by *shinkansen* (bullet train).

Takei and Nabeshima, who both previously worked at Tezuka Architects, came up with an ingenious approach that took advantage of the fact that the plot dipped below the level of the road winding around it. This meant that the ground floor, which is partially embedded in the hill, could be counted instead as a basement, allowing the architects to bypass the strict height restriction in the area and build upwards. As a result, Mount Asama came partially into view as the new house approached a height of 10 metres (33 ft).

Another aspect of the winning scheme made Ring House a showstopper. The mini-tower is structurally secured by twelve wooden pillars embedded in the concrete base, but these pillars recede behind striking horizontal bands of dark timber, which are covered with a veneer of scorched Japanese cedar and interspersed with clear glazing that wraps the tower. Not only did this arrangement give the resident a unique 360-degree panorama of the surrounding wood of pine, oak and chestnut trees, but also it made the tower float mysteriously during the day and glow at night.

The house is entered either via a bridge at the middle level or from the basement. All fittings and fixtures, including the cast-iron stove – an unexpected feature – are shaped to follow the exact dimensions of the pillars and glazing. Each floor offers 30 square metres (323 sq. ft) of open space, with shoulder-level partitions cleverly preserving the views of the trees beyond. Such stand-alone items as the dining-table were designed by the architects to complement the Wegner chairs and the light birch finish of the interior. Takei and Nabeshima say that they were careful to create furnishings with a clean finish, as the objects can be seen from outside at all angles through the bands of glazing.

The three-storey house includes a master suite with a sunken bath on the top floor, a loft-like kitchen and living room on the 'ground' floor and another bedroom in the basement. A *doma*, a gravelled area traditionally used for cooking, is incorporated into the basement, where wood for the stove is neatly stacked. The glazing becomes narrower at the top and in the basement to provide necessary privacy for the bedrooms. Otherwise, however, Ring House is no place for a recluse.

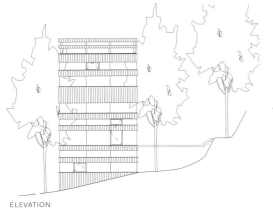
ELEVATION

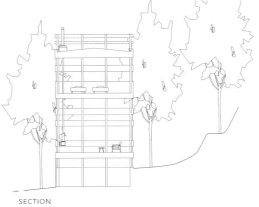
SECTION

BASEMENT, GROUND- AND FIRST-FLOOR PLANS

HOUSES AND HOUSING 219

Rooftecture S

Kobe, Hyogo Prefecture

ENDO SHUHEI ARCHITECT INSTITUTE, 2005

Shuhei Endo's first temptation on visiting the future site of Rooftecture S was to bolt. Thin and triangular, the lot was barely big enough for a two-person tent, let alone a full-sized house; in addition, there was a stone retaining wall on one side and a precipitous drop on the other. Much to Endo's chagrin, the clients, a retired couple, had already put money down, leaving him no choice but to make the attempt if he wanted the job.

Challenging as they were, these extreme conditions have given rise to a remarkable house. Draped with a folded rectangle of ribbed steel that forms both roof and façade, the house looks precariously balanced, yet five bulky pillars anchor it to the sliver of land. Equally impressive is Endo's inclusion of the surrounding scenery – a traditional method of visually extending a building's physical boundaries. Looking out over the railway tracks and motorway directly below, rectangular windows on one side of the house frame views of the Seto Inland Sea as far as the horizon. On the other side a wall of glass engages with the traditional rough-hewn blocks at the back, calling attention to the simple beauty of the mossy surface and turning the diamond pattern into an integral component of the interior.

The property belongs to an area on the edge of Kobe that was once dotted with beach homes but is today a quiet enclave for commuters, just a stone's throw from the railway station. A narrow road leads up to the lot, and stairs lead down to a wooden deck that nestles between the house and the wall. A transitional space reminiscent of an *engawa* (porch), it points the way to the front door.

As the clients requested, the upper floor is essentially one large room, with living and sleeping areas filling one half and an eat-in kitchen the other. The lower floor contains secondary functions, such as the *doma*, a multi-purpose area inspired by the dirt-floored vestibules of many old houses; a covered terrace; and the bath, appropriately a metal *Goemon-buro*, a type of tub steeped in history (named after a legendary Japanese hero – rather like Britain's Robin Hood – who was eventually boiled alive in a metal cauldron).

Despite the house's diminutive scale, the living quarters are unexpectedly roomy. Compensating for the shrinking floor area, the ceiling height rises rapidly as the room tapers to its 70-centimetre-wide (27½ in.) tip. An intimate 2 metres (6 ft) high in the sleeping area, it soars to 4.5 metres (14¾ ft) above the dining area. Perfectly sized for a table for two, the tiny space crescendos dramatically in a full-height window revealing a spectacular view of the suspension bridge to Awaji Island.

Sharp-edged and clad with steel, Rooftecture S required skilled workmen to bring it to life. But beneath this bold veneer is a homely abode that borrows subtly from tradition.

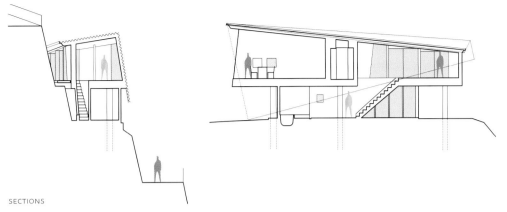

SECTIONS

OPPOSITE, TOP The house perches on a narrow strip of land. OPPOSITE, BOTTOM A folded steel plane forms the roof and part of the front façade. BELOW, LEFT The tapered plan culminates in the full-height kitchen window. BELOW, RIGHT The living area looks out on to the sea in one direction and a rough-hewn retaining wall in the other. BOTTOM LEFT The tip of the house measures just 70 centimetres (27½ in.); three poles jutting out at the bottom allow window cleaners to perch. BOTTOM RIGHT A railway line and a busy highway run directly below the site.

HOUSES AND HOUSING 221

Stage House
Karuizawa, Nagano Prefecture
TNA, 2006–2007

BELOW Perched on a steep cliff, Stage House opens to the forest with a cantilevered 'stage'. OPPOSITE, LEFT The house is entered from the roof. OPPOSITE, TOP RIGHT The tall glass façade allows views of the forest from every angle. OPPOSITE, CENTRE RIGHT The cascading interior space. OPPOSITE, BOTTOM CENTRE The change in level is kept just small enough to avoid the need for such safety measures as railings. OPPOSITE, BOTTOM RIGHT Everything in the house (apart from the timber) is in theatrical black and white.

Following the success of Ring House (pages 218–19), TNA's second house for a developer who owns an exclusive resort in Karuizawa turns an even more difficult piece of land into a sought-after weekend residence. Sitting on a steep slope and narrowly hemmed in by neighbours' plots, TNA's solution is enclosed with bands of steel that leave one great opening at the end, creating a 'stage' that offers an endless view of the forest, but no audience.

The new house is entered through a slit cut in the roof, reached by precipitous steps down the 7-metre (23-ft) drop from the main road. The cave-like entrance hall is flanked by a bedroom and a washroom, opening up to a generously proportioned space fitted with TNA's signature iron stove. While the steel roof points skywards, the floor follows the topography of the site down in terraces, and accordingly the ceiling eventually rises to a height of 9 metres (29½ ft), amply accommodating two levels: a lofty bedroom upstairs (reached by a floating steel staircase) and an open-plan living area downstairs. Steps descend the terraces to the kitchen, dining room and sunroom. Tall windows, more like transparent walls, slide to open the sunroom and jacuzzi at the tip of the house to the cantilevered timber terrace and the forest beyond, facing the sun, which is perhaps the only protagonist on this stage.

TNA's design language is gracefully simple, with just enough out-of-the-ordinariness to attract the sort of people who are looking for weekend homes in Karuizawa. Much to his delight, the developer managed to sell Stage House even before it was built, as was the case with Ring House. The lucidity of this building comes from the stepped, plateau-like profile of the ground floor, where differences in height mark out different areas. The 90-centimetre (35-in.) decrements are just below the height that legally requires the placement of such visually obstructive safety measures as railings. With storage space concealed below each platform, the space is kept free of clutter, and its minimalist economy is further enhanced by the seamless use of stained oak on the floor throughout.

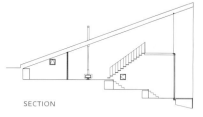
SECTION

FIRST-FLOOR PLAN

GROUND-FLOOR PLAN

ELEVATION

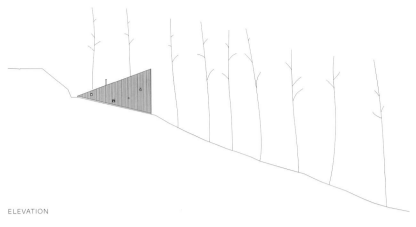

HOUSES AND HOUSING 223

Steel House

Bunkyo, Tokyo

KENGO KUMA & ASSOCIATES,
2004–2007

RIGHT, TOP A public ramp runs past the house; RIGHT, BOTTOM The gated main entrance. OPPOSITE, CLOCKWISE FROM TOP LEFT The garage and back door are accessed from above; the client's model train collection is displayed prominently in the living/dining area; a traditional-style tea room occupies the ground floor; stairs lead from the kitchen up to the back door.

In Tokyo, where – as we have seen – hopeful homeowners will build on just about any site, a skinny L-shaped lot encompassing a 6-metre (20-ft) drop is fair game. Although quirky, the site appealed to the clients, a family of three with two poodles, so they snapped it up and asked Kengo Kuma to create their dream house. When faced with these conditions, Kuma knew conventional construction just would not do. Turning his attention to retaining walls and bridges, he sought inspiration from civil engineering instead. His solution was a shell of corrugated steel plates that accommodates the site's peculiarities and saves space by serving as skin and structure in one.

Fortunately for Kuma, the family felt perfectly at home in a house built and shaped like a railway carriage. A professor at a nearby university, the husband is passionate about model trains, and display and storage space for his vast collection were the driving forces of the design. Another was the wife's desire for a traditional *tatami*-floored tea room.

Linking two different ground planes, the site faces a multi-lane thoroughfare lined with apartment blocks, but backs on to a quiet residential enclave at a higher level. The house may be entered through the garage above or the main gate below.

The front door opens on to a generous hall, spacious enough for the family ping-pong table and gracious enough to serve as the antechamber for the tea room beyond. From the hall, steps ascend to the family quarters. A single L-shaped space, the first floor holds the living and dining areas in one wing and the kitchen in the other. The second floor is also essentially one large room, consisting of sleeping berths and a bathroom plus a roof terrace overlooking the street. While the daughter's bed fills a nook at the back, her parents occupy an alcove to one side. Filmy white curtains serve as doors.

Considering it is encased in cold metal, the interior is surprisingly warm and welcoming. On the first floor, Kuma accomplished this with insulation and plastic wall panels that impart a greenish tint reminiscent of young bamboo. Built-in elements also soften the impact of the steel. While a 9-metre-long (29½ ft) showcase filled end-to-end with the professor's scale models dominates the living room, the tea room is the focus downstairs. A discrete place devoted to the ceremonial preparation and service of the steeped brew, the boxy space is defined by a wooden frame and enclosed with moveable screens made of coarse *washi* paper, but is open to a painted steel ceiling.

Part of the appeal of Kuma's architecture is the exceptional sensitivity and attention to detail in his use of traditional materials. But Steel House shows that he comfortably embraces modern, industrial materials, too.

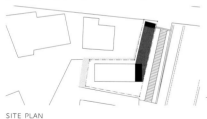

SITE PLAN

ELEVATION

HOUSES AND HOUSING 225

Tetsuka House

Setagaya, Tokyo

JOHN PAWSON, 2003–2005

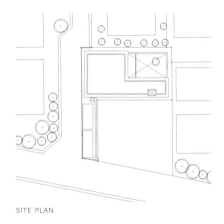

SITE PLAN

This small house in the quiet, leafy suburb of Setagaya – Tokyo's equivalent of London's Notting Hill – is pure John Pawson. Very well-known as a minimalist, Pawson has long mastered the art of fashioning pure, intense spaces of utter clarity: edges and joints are pristine, surfaces impeccably flat and spaces skilfully proportioned. This house contains very few spaces, but those that are present have been refined with great care. Plans of the building show even the precise positioning of furniture and the dimensions of the rugs.

Despite the four years he spent in Japan working for the esteemed interior designer Shiro Kuramata, Pawson's only opportunity for work in the country had heretofore lain in designing prestigious flagship stores for fashion houses Calvin Klein and Jigsaw. Tetsuka House is Pawson's first private house in Japan, and the project came through only because the architect took very seriously a handwritten letter that had arrived in his London office one day: a young couple asking if he would design for them a house just like the one used as the backdrop for pictures of food in the architect's cookbook, *Living and Eating* (2001). The house in question was Pawson's own house in Notting Hill.

Tetsuka House is approached via a long wall, which adds a sense of drama to the entrance. The ground floor of this box-shaped dwelling contains little more than a kitchen/diner, a traditional tea room and a small WC; there is also an external courtyard at this level. Stairs lead to a generous bedroom and a bathroom, both of which overlook the courtyard, itself a minimalist creation enlivened with a maple tree placed off-centre. Windows are imagined as frames of crafted views, rather than merely as devices for admitting light, while furniture is either integral or specially designed for the space it occupies. Plentiful storage is also built in.

Clad entirely in reinforced concrete, the building appears to be carved from a single solid block, complete with recesses for invisible light fittings. A long bench is also carved to emphasize the continuum from inside to outside. The kitchen column – with which one could almost spin the room – is the only rounded form, apart from the trunk of the maple in the courtyard; consequently, it creates an internal, artificial echo of the organic form outside. Where else, then, to meditate on this splendid oneness with nature but inside the 'sky shower' nestled next to the upper-level bathroom, where no ceiling prevents you from bathing in sun or rain?

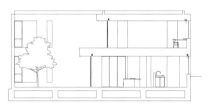

SECTION THROUGH HOUSE AND COURTYARD

ABOVE The lounge (LEFT) and traditional tea room (RIGHT) are seen from the courtyard. OPPOSITE, TOP LEFT The façade gives nothing away. OPPOSITE, TOP CENTRE The 'sky shower'. OPPOSITE, TOP RIGHT The kitchen column is the house's sole curved form. OPPOSITE, BOTTOM LEFT Furniture is simple and bespoke. OPPOSITE, BOTTOM RIGHT A slot of coloured light animates the whiteness of the stairway.

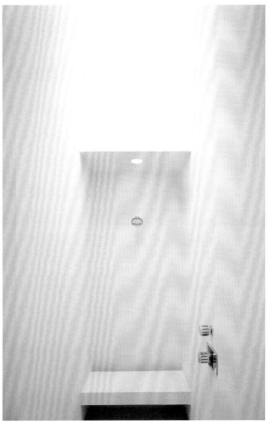

Y House

Chita, Aichi Prefecture

KEI'ICHI IRIE + POWER UNIT STUDIO, 2002–2003

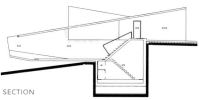
SECTION

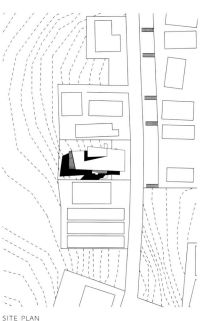
SITE PLAN

Clad entirely in reinforced concrete, and looking almost like a suit of armour, Y House stands defiantly on a hilltop in an affluent suburb of Nagoya City. A war has indeed been waged by the maverick designer Kei'ichi Irie to combat the 'sad' state of Japan's suburbs, which are full of lifeless, pre-packaged houses.

Surprisingly modest, Y House reveals little from the street. Eager to screen off unsightly views, the architect wrapped the house with angular, gravity-defying concrete slabs, which – thin as they are – give the impression of buoyancy with no visible means of support. Only inside is the house's full force deployed: a dark, tunnel-like living room follows the site's topographical slope with a steep, exposed-concrete floor, leading to the back of the house and ending in an explosion of natural light on a cantilevered balcony, which extends into the garden and overlooks a wood.

Privacy is maximized, and curtains and blinds are put away. The kitchen floats on one side of the living room, and the bathroom on the other.

The visual acrobatics and bunker theme continue with the clients' studio, an observation pod that overhangs the living room. Its large glass screen offers a vantage point for observing the comings and goings in the room below, adding a curious contour to the house. A small rectangular hole discreetly inserted at the corner of the studio offers another – more secretive – vantage point, revealing the basement roughly 10 metres (33 ft) below. Despite having no windows, thanks to this device the underground space feels airy.

The clients, a young couple who both work in the creative industry, engaged fully in a dialogue with the architect and his construction team all the way through the project's

making. Their full support, it seems, enabled the architect to carry out the apparently impossible task of fabricating a house with load-bearing concrete merely 15 centimetres (6 in.) thick and bent at various acute angles. Such technical complexity has resulted in some imperfections and idiosyncrasies, but without them the house – both daring and playful – would be a less characterful achievement.

OPPOSITE, TOP Architect Kei'ichi Irie combats the 'sad' state of Japanese suburbs by wrapping the house with gravity-defying concrete slabs. OPPOSITE, BOTTOM The young clients enjoy the forest view. RIGHT Street view. FAR RIGHT The bathroom. BELOW AND BOTTOM The house descends sharply from street level to the living area, while the study hovers above. BOTTOM RIGHT The spacious basement bedroom.

Yakisugi House

Nagano City, Nagano Prefecture

TERUNOBU FUJIMORI, 2005–2007

RIGHT, TOP The tower contains a tea room. RIGHT, BOTTOM *Yakisugi* means 'charred cedar wood'. OPPOSITE, TOP LEFT The periscope, which gives the Buddha statue (visible in the image opposite, bottom right) a halo of light, is below the tower. OPPOSITE, TOP RIGHT The cave-like living room looks on to the garden and storehouse. OPPOSITE, BOTTOM LEFT Pieces of charcoal are stuck like a swarm of bees around the stove. OPPOSITE, BOTTOM RIGHT The light, the dining table and chairs and the firewood container by the stove are all designed by Fujimori.

The idea for Yakisugi House reflects Terunobu Fujimori's historical insight (he is an architectural historian as well as an architect): its inspiration came from a small cave dwelling near the Lascaux caves in France, which he recently visited. A retired couple, whose family has lived on the same plot of land for generations, decided that their old house, built in the late Meiji period (1868–1912), had become a burden to maintain and live in, and commissioned Fujimori in 2005 to build them a new home.

The cave idea materialized as the main living and dining room, which leads to a study, two bedrooms and a tea room inside a mini-tower. Fujimori then wrapped his 'cave' with charred cedar boards: a traditional and durable cladding material still used in Okayama Prefecture in the south of Japan, and reflected in the house's Japanese name. A big outdoor party was organized to entice the architect's friends and those of the clients themselves to char the 8-metre-long (26¼ ft) boards. The boards are traditionally available in lengths of less than 2 metres (6 ft), for if they are any longer they warp with the heat of the production process, but here, the inevitable warping of the longer boards was remedied by filling the gaps with thick mortar, creating the striking zebra pattern of the exterior walls.

The architect also landscaped the generous 1825 square metres (19,650 sq. ft) of land, planting his bow-shaped 'green sculpture' beside a small hut that serves as a rest house, reviving the old well and directing fresh spring water from it through a long bamboo conduit. Finally, he paved the walkways, creating a textured surface by using brush strokes on wet mortar – a technique he developed himself.

Three tree trunks penetrating the end of the gently sloping roof appear to have some ritualistic significance, but the architect says he simply wanted to continue the three supporting columns skywards. Light brought in through the small, periscope-like chimney attached to the opposite end of the building, on the other hand, enhances the religiosity of the small statue of Buddha, carved out of a leftover piece from the free-standing mulberry pillar that supports the ceiling of the main room.

Through elements ranging from the hand-rolled copper plates covering the roof of the tower to pieces of charcoal stuck like a swarm of bees around the stove, Fujimori invariably – and sometimes idiosyncratically – confronts the functionalist/minimalist language of modern architecture. His contemporary Kengo Kuma has called his approach 'anti-historical'; one could also call it 'anti-geographical'. It would be very difficult indeed to attribute the overall style of Yakisugi House to any specific place or period.

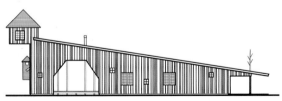 

ELEVATIONS

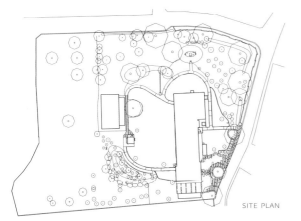
SITE PLAN

HOUSES AND HOUSING 231

- 234 Aoba-Tei
- 236 Dior Omotesando Store
- 238 F-Town
- 240 Flower Shop H
- 242 Gyre
- 244 hh Style
- 246 Honmura Lounge and Archive/Benesse Art Site Naoshima
- 248 House and Office for Atelier Bow-Wow
- 250 MIKIMOTO Ginza 2
- 252 One Omotesando
- 256 Prada Aoyama
- 258 Shin-Marunouchi Tower
- 260 Showroom H
- 262 Tod's Omotesando Building
- 264 Toririn
- 266 Triad

Offices and Retail

Aoba-Tei

Sendai, Miyagi Prefecture

ATELIER HITOSHI ABE, 2003–2005

Set diagonally opposite Toyo Ito's Sendai Mediatheque (pages 84–85), on one of Sendai's main streets, is the exclusive French restaurant Aoba-Tei. Hitoshi Abe, whose office is based in this hip northern city, was commissioned in 2003 to design the interior of the restaurant, which is on the ground and first floors of a seven-storey commercial building.

Abe created a second layer of walls from thin steel sheets, which were moulded in a factory but welded on site to create a seamless structure. The extra layer not only completely transforms the interior but also functions as a soft interface linking what is inside to what is outside, since the patterns on it refer to the zelkova trees planted in a line at the front of the building.

The restaurant is open only by request and then only at night, so it is fitting that the new interior comes alive when it is dark outside. The cave-like inner walls are punctured with hundreds of thousands of 'graphic holes' in three different sizes – 4, 6 and 9 millimetres (⅛–⅓ in.) in diameter – so that the illuminated images of the trees are revealed through them.

Restaurants in Japan are often brightly lit to draw in customers; Aoba-Tei, then, is unusually subtle. The ground floor holds a reception area, a cloakroom and a floating staircase animated with the restaurant's only fluorescent light. The steel interface rises out of the building's concrete base, up the walls and ceiling of the ground level and over the walls, part of the floor and the entire ceiling upstairs. The horizontal division between the ground and first floors is almost nullified.

Although the steel mechanical rods supporting the interface are visible from the outside, by the time we climb the stairs and arrive on the first floor to sit at the bar counter or one of the tables, we are transported into an abstract version of the real world, still visible from inside the restaurant. Our own physical bodies seem also to turn into a series of dots as we digest what surrounds us. The restaurant perfectly complements the Mediatheque, which also embraces the virtual.

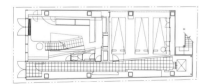

FIRST-FLOOR PLAN

GROUND-FLOOR PLAN

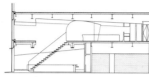

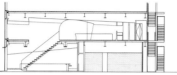

SECTIONS

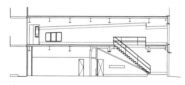

ELEVATION

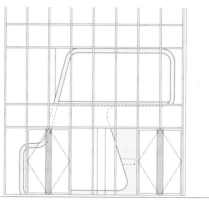

ABOVE View of the restaurant from the street. OPPOSITE, LEFT, TOP AND CENTRE The restaurant and bar on the first floor. OPPOSITE, TOP RIGHT The stairway becomes the source of light in a darkened space. OPPOSITE, BOTTOM LEFT The reception area on the ground floor. OPPOSITE, BOTTOM RIGHT The 2.3-millimetre-thick (¹⁄₁₀ in.) inner walls are perforated with hundreds of small holes of varying sizes, so that when the steel 'skin' is backlit, the image of overlapping tree branches appears.

OFFICES AND RETAIL 235

Dior Omotesando Store

Shibuya, Tokyo

KAZUYO SEJIMA + RYUE NISHIZAWA/SANAA, 2001–2003

This store for Christian Dior, located on Omotesando Avenue (Tokyo's prime shopping street) close to such landmarks as Tod's by Toyo Ito (pages 262–63) and Prada Aoyama by Herzog & de Meuron (pages 256–57), is a cool, glassy, translucent box that barely hints at what is inside. Conceived to have a deliberate mystique, it has a double skin that hides rather than reveals its content.

The almost seamless external glazing is reminiscent of Toyo Ito's Sendai Mediatheque (pages 84–85; Sejima worked in Ito's office before setting up SANAA), but the inner layer of folded, translucent acrylic sheets gives the Dior building a softer appearance, more suited to the world of fashion. At night it takes on an ethereal glow. The acrylic sheets – which filter the light and give the building its glacial sheen – are printed with white lines, and the resulting moiré intensifies the sense of otherworldliness.

Most of this tall building – at 30 metres (98½ ft), it is about as high as any building is allowed to be in this built-up area – is given over to retail accommodation; there is a multi-purpose event space on the fourth floor, above which is a rooftop garden. The crisp modernity of the façade gives way to the Dior design team's eclectic interior fit-out of ornate mouldings, parquet flooring and timber panelling, developed in collaboration with various artists.

Architectural critics have often commented that SANAA's work seems to be completed only by habitation; in other words, it is the inhabitants and their trappings that give the buildings movement, colour, texture and life. The spaces themselves are

simple, make the most of natural light and seek only to serve their purpose. This is certainly the case for the practice's New Museum in New York, and the Dior building follows similar principles, although its impact did not depend upon the arrival of shoppers. Instead, the fashion house's own designers provided a ready-made 'personality' for the interiors.

OPPOSITE Glazing and acrylic sheets lend the building an ethereal glow. TOP LEFT Located on Omotesando Avenue, the store competes with prestigious neighbours for attention. CENTRE LEFT The façades present an impression of varying floor heights. RIGHT A cool, glassy box, the building barely hints at what is inside. BOTTOM LEFT The building reaches as high as regulations permit.

OFFICES AND RETAIL 237

F-Town

Sendai, Miyagi Prefecture

ATELIER HITOSHI ABE, 2003–2007

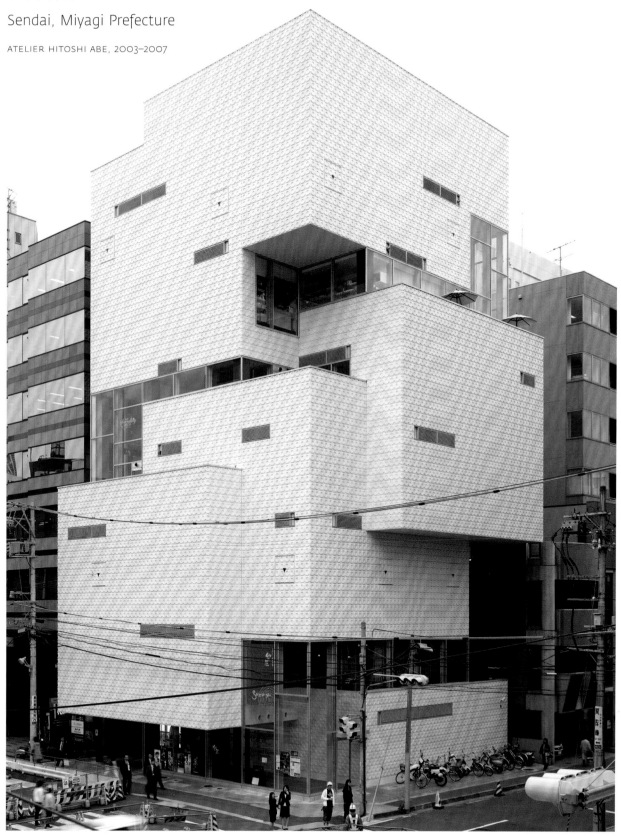

All over Japan, where most people socialize outside their homes, multi-storey eat-and-drink buildings abound. Clad with a vertical strip of signage that reads like a table of contents, each is loaded with small bars and restaurants. When asked to design a building of this ilk for a corner site in his home town of Sendai, Hitoshi Abe recognized the chance to rethink not just the appearance but also the inner workings of such a formulaic building type.

The typical Japanese restaurant tower offers a wide choice of food, ranging from sushi to spaghetti, with independent eateries stacked one on top of the other. F-Town, by contrast, is a spiralling sequence of tenant spaces contained within an assemblage of six cubes articulated on one or both of the building's street fronts.

The double-height boxes conceal seven storeys of separate spaces suitable for restaurant use. As in conventional bar buildings, shared lifts and stairs lead to each level, but voids between the boxes offer the potential for inserting additional steps to create two-storey cafes or to connect individual units internally. If exercised by tenants, these space-planning options could enable their customers to move effortlessly from place to place, much as they do outside.

From the street, F-Town stands out dramatically from its surroundings, a burgeoning entertainment district between the city's main railway station and its baseball stadium. In addition to its unusual massing, the building is clad entirely with white panels that look like expensive tiles but are made of low-cost, lightweight concrete. There is no need for attention-grabbing

OPPOSITE The seven-storey building is composed of a stack of boxes. BELOW The exterior cladding has the appearance of tile but is actually low-cost, lightweight concrete panels.

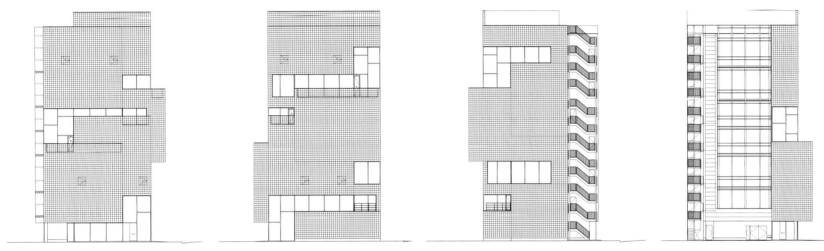

ELEVATIONS

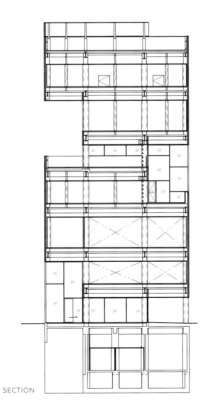

SECTION

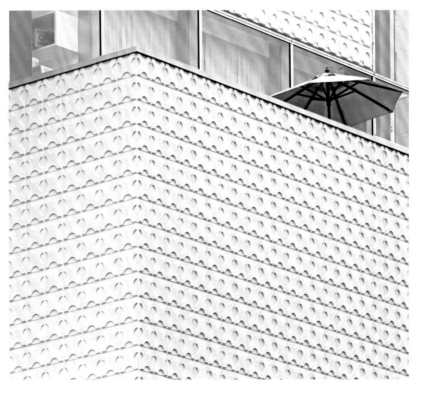

neon here. Created in collaboration with the graphic designer Asao Tokolo, the panels are intricately patterned with carved lines produced by a digital milling machine. From a distance, the two patterns blur together, but on closer inspection it can be seen that one is composed of circles and the other crosses. Covering the surface of each box with blocks of the same pattern reinforces the building's cubic massing. And when the sun sets, shadow underscores the rich texture and brings the building to life.

Because the interior design of each unit was beyond the scope of Abe's commission, the individual restaurateurs can decide whether to connect to their neighbours. But there is strength in numbers: if they choose to realize the architect's vision in full, the unbroken chain of bars and cafes is bound to attract more customers and turn F-Town into an entertainment destination in its own right.

OFFICES AND RETAIL 239

Flower Shop H

Tokyo

OFFICE OF KUMIKO INUI, 2005–2009

Most park buildings pale in comparison with their lush natural surroundings, laments Tokyo architect Kumiko Inui. Her competition-winning proposal for a flower shop in a central Tokyo public garden, however, raises the standard to an entirely new level. A replacement for a 1970s store that was no longer sufficiently earthquake-resistant, the project is a cluster of five small but tall buildings that evoke the atmosphere of the outdoors despite being encased in glass and granite.

Inui's intention was neither to mimic nature nor to dissolve the boundary between interior and exterior – her goal was far subtler. Through light, height and transparency she hoped to evoke the ambiguity of enclosure and exposure created by the tree canopy. Instead of muscling her way in with a single, solid mass, Inui envisaged the diminutive pavilions coexisting amicably with the setting by nodding politely to the architecture outside the park and without overwhelming the landscape inside.

Divided by function, Inui's structures are essentially free-standing rooms consisting of a sales area, a customer meeting room, a flower bar, a workroom and a service core (plus an enclosed mechanical area), all unified by a rammed-earth terrace. Spiral stairs lead down to the office and another work area underground; there is also a rammed-earth terrace at street level.

Although the proprietor wanted a bigger shop, the architect could not legally exceed the area or height of the original store, which was an exceptional 7.5 metres (24⅛ ft) tall. Despite this generous vertical dimension, Inui had to stick to a single-storey scheme above ground, but within the site's boundaries she could lay out the horizontal plan as she wished. Separated by narrow gaps, most of which are less than 1 metre (3 ft), the independent components make the shop seem larger without covering more ground.

While the interstices are dark, the interiors are soaring and filled with light thanks to the huge glass panes integral to each. As well as emphasizing the buildings' remarkable stature, the transparent sheets form the thinnest of barriers between inside and out. Most are fixed – strong breezes are unkind to delicate blossoms – but some are automatic doors, their movement carefully choreographed to allow smooth passage between pavilions. The only direct connection, unsurprisingly a transparent tunnel, is between the flower bar and the work area.

Secured by steel frames, the boxes are structurally independent, too. Since the ground-floor and basement levels are not congruent – the ground floor holds the individual boxes independent of the rooms in the basement – unusual beams were needed to tie the entire shop's structural system together. But these manoeuvres are coolly concealed by granite veneer and plasterboard.

More substantial than a simple greenhouse but less robust than a conventional stone-clad edifice, Flower Shop H blends beautifully with the park and its seasonal offerings.

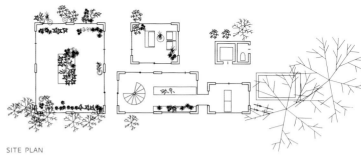

SITE PLAN

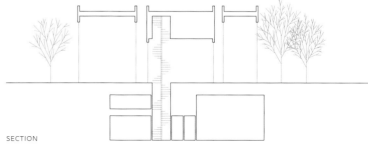

SECTION

OPPOSITE, TOP Exceptionally tall window walls blur the boundary between inside and out.
OPPOSITE, BOTTOM The shop consists of a group of buildings.
LEFT, TOP The granite-veneer exterior cladding recalls the commercial buildings near by.
LEFT, BOTTOM The white plasterboard interior finish is the perfect backdrop for the colourful flowers.

Gyre

Shibuya, Tokyo

MVRDV, 2004–2007

SITE PLAN

When the Dutch practice MVRDV, well known for its thoughtful, research-based designs, was commissioned to design a shopping centre on Tokyo's prestigious retail avenue, it reacted against the deliberately iconic stores that have recently arrived in the area. These buildings, reasoned MVRDV, were the architectural equivalent of supermodels: although beautiful, with almost everything invested in their external appearance, they could also be perceived as exclusive and intimidating. MVRDV set out to create instead something where form – not just skin – would become the defining characteristic.

The design team eventually created a distinctive multi-level structure not by wrapping it in an eye-catching skin but by breaking it down into distinct units. The building remains part of the public realm, while its levels are alternately pushed and pulled to create a complex form of cantilevers, acute angles, overhangs and cutaways. With its encircling stairs, the building resembles a large rock formation, turning idle shoppers into chic urban rock climbers.

'A building in the shape of a gyre is a body in motion,' explains the centre's website. The complex form radiates from a central core, each storey twisted to create terraces at intermittent levels. The architects focused on the vertical movement of people – the spatial quality of which was also explored in Maki and Associates' splendid Spiral building of 1985 on nearby Aoyama Dori – and it was through this massing exercise that the character of the building emerged. 'As the silhouette of the building is already unique, the façade can stay relatively modest, allowing the occupants of the spaces to express themselves,' the architects say.

Originally, escalators were to be on the edge of the building, but eventually they were moved to the building's core. Inside this concrete-and-steel building of seven storeys (including two basement levels), such big brands as Chanel and Bulgari sit side-by-side with less well-known retailers. As is usual, the fit-out was left to the design teams selected by the tenants.

MVRDV has survived the fiercely competitive architectural catwalk of Omotesando by observing the trend and reversing it. For example, the architects have flipped the circulatory structure designed by Tadao Ando for Omotesando Hills, the new shopping centre opposite Gyre, with a swirl of walkways inside a more conventional glass box that tapers at one end to fit a triangular plot. By going under the skin of the building, MVRDV has explored form in visceral ways, stepping away from the mindless strolling induced by most shopping malls.

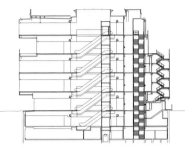
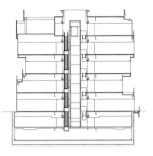

SECTIONS

OPPOSITE, TOP The building is an assembly of cantilevers, acute angles and cutaways. OPPOSITE, BOTTOM Dusk view. LEFT The architects have emphasized the silhouette, allowing the façade to remain modest. BELOW, LEFT The building's horizontal elements twist, providing space for external circulation. BELOW, RIGHT Big brands occupy the building alongside less well-known retailers. BOTTOM The escalators once planned for the exterior are now at the core of the building.

OFFICES AND RETAIL 243

hh Style

Shibuya, Tokyo

TADAO ANDO ARCHITECT & ASSOCIATES, 2004–2005

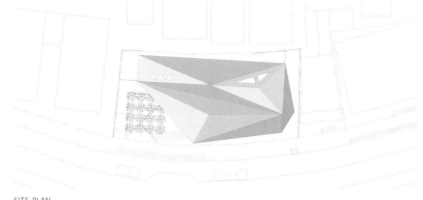

SITE PLAN

Down a small alley just a few steps from hh Style's flagship store – a semi-translucent glass box designed a few years previously by the Tokyo-based practice SANAA – Tadao Ando's robust store for the furniture-import company is almost windowless, save a long, narrow horizontal slot at eye level. Clad in dark steel, Ando's new building seems to be in direct confrontation with SANAA's original building; the Dutch practice MVRDV similarly responded to the Japanese duo's ethereal Dior building on Omotesando (pages 236–37) by turning its Gyre shopping centre just a few doors down into a dark mass (pages 242–43).

This turning away from the busy street to create an inner sanctum is a feature of Ando's designs. A series of shadow studies helped determine the shop's relatively compact, faceted exterior. The steel plates, which are just 16 millimetres (⅔ in.) thick, are folded rather than bolted or riveted, and reflect light at different angles, elegantly hinting at the obverse and reverse of origami papers and giving a smooth, origami-like effect of great lightness and whimsicality.

The faceted, 'invisible to radar' envelope provides a foil for the orthogonal concrete floor plates, columns and stairs. The strict height restriction did not stop the architect from digging deep to create a surprisingly large space, and the transition from the smooth, dark exterior to the ribbed, white interior is particularly striking. While the dark timber flooring offsets the white-painted steel, the smooth, angular cuts are echoed everywhere, from the pitched entranceway to the sharp voids made in the concrete walls, as well as the slanted corrugated walls and the ceiling that culminates dramatically in a single skylight, positioned neatly inside a triangular plate.

Each space has its own distinct character while forming part of a coherent, integrated whole. Ando's faceted, muted treatment of exterior and interior suits the brief, which was to house a new furniture range by the Italian fashion designer Giorgio Armani. The project was a departure for Ando, who had previously explored a language of flat and curved surfaces in concrete, but he is clearly happy to take on a new challenge, as his deft handling of a site with so many constraints demonstrates.

ABOVE, LEFT The building is almost windowless, except for a narrow horizontal slot. ABOVE, RIGHT Steel plates appear to be folded, origami-like, from a single sheet. OPPOSITE, TOP LEFT Interior volumes are pinned around a concrete core. OPPOSITE, BOTTOM LEFT The spaces retain their own characters, while still forming part of a coherent whole. OPPOSITE, RIGHT The corrugated interior surfaces contrast with the 'stealth' aesthetic of the exterior.

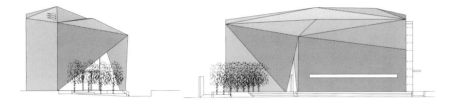

ELEVATIONS

OFFICES AND RETAIL 245

Honmura Lounge and Archive/ Benesse Art Site Naoshima

Naoshima, Kagawa Prefecture

OFFICE OF RYUE NISHIZAWA, 2003–2004

This small island between Shikoku and Honshu has been given a new lease of life as a major centre of contemporary art, thanks to a far-reaching vision that Soichiro Fukutake, president of the Naoshima Fukutake Art Museum Foundation, had in 1989. The Osaka-based architect Tadao Ando designed significant elements of this dispersed project, including Benesse House and the largely underground Chichu Art Museum (pages 64–65), while the Tokyo-based practice SANAA embarked on the island's chic entranceway, Naoshima Ferry Terminal (pages 49–49). A score of artists and architects have been allowed to add, reuse and mould existing buildings (including a 200-year-old *minka* or traditional Japanese house), and even to create a new building for art installations in the village of Honmura.

The Honmura Lounge and Archive is the result of a similar intervention by architect Ryue Nishizawa, one half of the practice SANAA – only this time, the purpose was to create a functioning office space for the staff and an information resource and sales point for the wider Naoshima art site. Nishizawa's bold yet subtle twist on a former Nokyo supermarket (an agricultural cooperative) is, however, just as stimulating and refreshing as the previous works for the site.

Like the old butcher who deftly chops a carcass into different parts in an ancient Taoist tale, Nishizawa removed great chunks of wall, floor and ceiling with ease and confidence to create spacious rooms filled with light. The steel-and-concrete structure of the original building is preserved to give a rough-and-ready aesthetic of welds, bolts and sliced-through floor slabs. That roughness, along with the original nondescript 1970s windows and ceiling, forms a striking contrast with the slick metal railings and stairways, lightweight furniture and soft textiles. Daylight is brought deep into the double-height space through glazing in the pitched roof, while glare is reduced by a fabric canopy.

For a country that habitually demolishes and replaces buildings deemed to be of low historical or cultural value, this project is a remarkably sensitive and intelligent reuse of a bland and straightforward structure. It is a poignant example of recycling and the result is a building with character, both raw and refined. The difference between the new and the old is clear, generally expressed through rough and smooth surfaces. With increasing concerns for the environment, we should be seeing more interventions like this in Japan.

OPPOSITE, LEFT The building is an elegant example of reuse and refurbishment. OPPOSITE, RIGHT A fabric canopy modulates the light. BELOW The building's former incarnation as a supermarket is unsuspected; its upper level translates easily into office space. BOTTOM LEFT Rough cuts and welds are clearly visible. BOTTOM RIGHT Floor plates and roof slabs are cut through to create a light-filled double-height void at the back.

House and Office for Atelier Bow-Wow
Shinjuku, Tokyo

ATELIER BOW-WOW, 2005

BELOW, LEFT Large windows overlook the congested context. BELOW, RIGHT Occupying a flag-shaped plot, the house is tethered to the street by a narrow walk just wide enough for an emergency vehicle. OPPOSITE, TOP Two views of the office work space. OPPOSITE, BOTTOM, LEFT AND CENTRE More than way stations, stair landings double as such functional areas as the kitchen and cosy living room. OPPOSITE, BOTTOM RIGHT Part of the office by day and the architects' home by night, the kitchen marks the transition between public and private realms.

In a city as densely packed with detached houses as Tokyo, gaps, glitches and odd geometry are inevitable. These leftover lots are marginalized by most, but Atelier Bow-Wow values them as components of the urban fabric.

Unsurprisingly, when the firm's husband-and-wife principals decided to consolidate their home and office, they chose a flag-shaped site that would scare off many prospective buyers. Although inexpensive and centrally located, the modest parcel was tightly squeezed between buildings and completely cut off from the street but for two narrow passages.

Flag-shaped sites are increasingly common in Tokyo, and result from the division in two of a rectangular property, since all residential lots must by law be directly accessible from the street for emergency vehicles. Some designers respond to these awkward, cramped conditions with introverted homes that ignore their surroundings, but Atelier Bow-Wow was inspired by the snippets of views, hidden crevices, uncultivated greenery and even blank walls to be seen from the property.

After studying sight lines and taking stock of the context, the architects turned their attention to the building's inner workings. Because the space needs for home and office fluctuate during the week (and sometimes even over the course of a single day), they opted for a continuous space over four storeys, gradually shifting from public to private.

The sequence begins with the basement studio, followed by the foyer and administrative offices on the ground floor. Used by both the couple and their staff, the first floor holds the kitchen and living room, with a large table that easily adapts to meetings and late-night model-making sessions. The bedroom fills the second floor; above is a roof terrace lined with potted plants.

Sliced vertically as well as horizontally, one-third of the floor area is given over to stairs – an unusually high proportion for a building with such a small footprint. But Atelier Bow-Wow's steps are not just floor-to-floor conveyances. While the open treads provide a gentle transition between levels, the oversized landings act as rooms, each – foyer, kitchen and study, for example – with a distinct character and defined purpose.

In a similar vein, the architects turned exterior 'deficits' into design assets. Instead of editing the mottled stucco wall only inches away, they incorporated it: fronted by full-height glass, it acts as wallpaper. Instead of shunning the haphazard roofscape at the back, they embraced it with a loggia-style balcony. And instead of disregarding the nondescript house next door, they aligned their interior columns with its outer wall, firmly anchoring their building in its broader context. Orientated outwards, this building is at one with its immediate environs and remarkably at peace with the prevailing crowded urban condition.

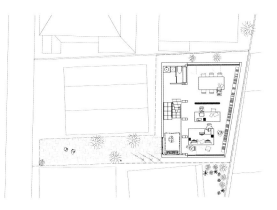

SITE PLAN

BASEMENT PLAN

FIRST-FLOOR PLAN

SECOND-FLOOR PLAN

ROOFTOP PLAN

OFFICES AND RETAIL 249

MIKIMOTO Ginza 2

Chuo, Tokyo

TOYO ITO & ASSOCIATES, ARCHITECTS
2003–2005

LEFT A shadow gap separates inhabited space from rooftop machinery. RIGHT Irregularly shaped windows suggest the forms of natural pearls. OPPOSITE, TOP LEFT Inside, too, the building is a combination of flat surfaces and gentle curves. OPPOSITE, TOP RIGHT The organic forms of the windows are picked up in the sweep of the feature staircase. OPPOSITE, BOTTOM The façades are composed of a thin sandwich of steel and concrete – both decorative and load-bearing.

The famous jewellery store MIKIMOTO – founded in 1899 by Kokichi Mikimoto, a pioneer in cultured pearls – commissioned Toyo Ito to design its second building in upmarket Ginza. The first is positioned at the heart of the district, 1 square metre (10¾ sq. ft) of which was once heralded as the most expensive plot of land in the world. The area is peppered with an assortment of stylized buildings, from neo-Gothic and neo-Tudor, to the art deco Kojunsha Club (Japan's first gentleman's club, built in 1929 and recently refurbished) and the neo-Renaissance Wako building of 1932 with its landmark clock, as well as such recent additions as Jun Aoki's Louis Vuitton and Renzo Piano's Hermès stores.

Ito's new nine-storey building, a launch pad for the retailer's new range of cosmetics, is wrapped in a pink skin punctured with 163 irregular holes, which appear to glide through the surface as if they were bubbles rising from submerged oyster shells. At first distracted by such an illusion, the viewer does not notice that this is a structurally daring building that stands as an excellent example of collaboration between architect and engineer. Developed with engineer Mutsuro Sasaki, the tower has no columns – the four sides of the box are both surface *and* structure.

The façades are constructed of a super-thin sandwich of steel and concrete: 12-millimetre-thick (½ in.) steel plates filled with 200 millimetres (nearly 8 in.) of concrete. The building is effectively a giant tube, into which openings are cut almost randomly, such is the stiffness and strength of the structure. Once the plates were in place, the welds were ground down and the steel was rust-proofed and painted. In a conversation between Ito and Sasaki recorded for the publication *GA Japan* in 2006, both men admit to being equally fascinated by the sweeps and free-form curves of Antonio Gaudí, and the rectilinear control of Ludwig Mies van der Rohe. Ito's tower, then, embodies two contrasting styles of architecture in one stroke.

An oddly fascinating feature of this building is that the façade openings do not correspond with the floor plates. A cursory look at the façade would suggest that the walls could not possibly be load-bearing, that they must surely be hung purely as a decorative skin; but internally, the lack of columns belies that impression. The visibility of the floor plates is surely an important part of that architectural conceit – or perhaps the point of this exercise, other than its function as an elegant shop and restaurant, is its ambiguity.

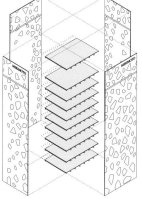

EXPLODED AXONOMETRIC

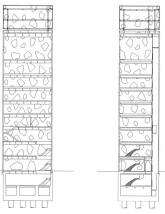

SECTIONS, WITH WINDOW CONFIGURATION

OFFICES AND RETAIL 251

One Omotesando

Shibuya, Tokyo

KENGO KUMA & ASSOCIATES, 2001–2003

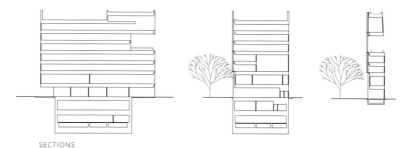

SECTIONS

BELOW, LEFT The façade is veiled in vertical strips of timber. BELOW, RIGHT Dusk view. OPPOSITE View from the cantilevered boardroom at the top of the building. OVERLEAF The reception furniture, all designed by Kuma, becomes a light-source.

Positioned at one of Tokyo's biggest intersections – where Omotesando, a tree-lined street culminating in Meiji Shrine, crosses another large avenue, Aoyama Dori – One Omotesando is the head office and retail centre for fashion group Louis Vuitton. The architect Kengo Kuma decided to clad most of its façade with vertical strips of timber, a material that is normally banned from the exteriors of buildings in large urban areas because of the fire risk. However, negotiations with the city authorities and the inclusion of fire-safety measures enabled Kuma's passion for incorporating something old into the new to be accommodated.

'In the past Tokyo was a city of wooden architecture. The human scale of wooden buildings and the warmth and softness of this natural material made Tokyo uniquely appealing,' Kuma says. His decision to employ 45-centimetre-deep (17¾ in.) mullions made of larch fins also reflects its surroundings, emphasizing the verticality of the zelkova trees; the timber's function as a carbon sink (locking up carbon dioxide until it burns or is decayed) also reflects the larger context, in which our environment is of paramount importance. In addition, the mullions act as sun shades: by admitting sunlight only when it strikes the façade of the building directly, they passively conserve energy, reducing the need for artificial cooling.

The result is that the solidity of the building changes according to the lighting conditions and the angle from which it is viewed. When the lights are on inside, for example, the façade all but disappears. This building amounts to more than just a pretty façade, however. Volumetrically, too, it is interesting – especially the upper-level boardroom, which cantilevers over a void as if held in place by nothing but filigree timbers.

Inside, the themes of lightness, transparency and verticality continue. Furniture in the reception area is lit brightly from within, while screens and stairways are reduced to fine lines and pure forms. Artificial lighting is delivered through channels and translucent materials so as to wash down in controlled beams or a subtle glow. The interior palette is a neutral one of greys, whites and timber; it is the treatment of light that gives the building its warmth.

SITE PLAN

OFFICES AND RETAIL 253

Prada Aoyama

Shibuya, Tokyo

HERZOG & DE MEURON, 2000–2003

RIGHT, TOP Setting the building back from the street and creating an empty, plaza-like space on the plot, Herzog & de Meuron evokes a sense of luxury. RIGHT, BOTTOM Complex zoning laws dictated the shape of the building. OPPOSITE, CLOCKWISE FROM TOP LEFT The internal spaces defy simple description; views are distorted through the bubble-like glass; futuristic information monitors snake through the space; a glimpse of Tokyo's skyline from a higher level.

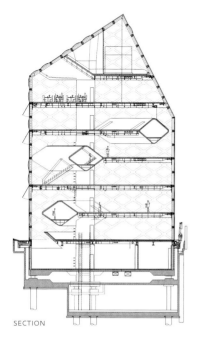

SECTION

Herzog & de Meuron's scheme for Prada was envisioned as one of the fashion house's 'Epicentres': a series of stores designed by architectural titans and intended to be much more than just a shop. The store sits on a relatively tight plot in Tokyo's Aoyama district, surrounded by a jumble of undistinguished buildings rarely more than four storeys high, occupying every inch of available land. This context gave the practice the excuse it needed to do something daring: 'We didn't want the building to sit on the street in the same squat and stocky way as its neighbours,' said Jacques Herzog. 'In addition, we wanted to evoke a sense of luxury and create the kind of public space often seen in Europe, which meant not building on part of the land.'

At six storeys, the shop is taller than anything else in the immediate vicinity, and highly visible. Herzog & de Meuron also worked with the complex zoning laws, which fill the air with an invisible matrix of lines and restrictions; by moving the notional building around the site, they were able to change its form in response to the web of planning parameters (which are set out like architectural ley lines).

The result is a crystalline form on the outermost corner of the site, clad in curved glass panels as though enshrined in bubble wrap. A sunken entrance bends around a corner, only to resurface curiously like a trumpet with a square bell at the opposite corner. This solution creates a small pocket of air, introducing European gracefulness to a city starved of open space.

The architects resisted the temptation to do the usual thing and create a 'fat, load-bearing core' around which the rest of the building would be built; instead, a lattice of large tubes, providing circulation routes and semi-intimate spaces, supports the structure. The store is then wrapped in a grid of steel and glass, in much the same way that a string bag wraps its contents with efficient ease.

The internal spaces created by such an exercise defy simple explanation. The building's prism-like structure yields distinctive interiors that frame the outside world through glazing units that are variously convex, concave and flat. The architects were responsible for the interior design, including all furniture, and a restrained palette of materials and finishes, including cream-coloured carpet and leather, timber flooring and perforated metal sheeting, unifies – indeed homogenizes – the spaces. The building is impeccably detailed, with façade and interior work forming a single, almost seamless object.

OFFICES AND RETAIL 257

Shin-Marunouchi Tower

Chiyoda, Tokyo

HOPKINS ARCHITECTS, 2001–2007

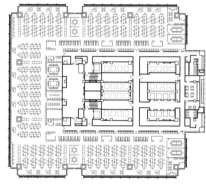

TYPICAL FLOOR PLAN

SITE PLAN

The Shin-Marunouchi Tower, opposite Tokyo Station, is the tallest building in the city's financial district. At 198 metres (650 ft) high, the tower is not large by international standards, but this part of the capital is carefully controlled by planning and zoning laws, which restrict the heights of buildings so that the nearby Imperial Palace does not become overlooked. The British firm Hopkins Architects was given the commission, since it was deemed that, being British, it would respect the historic nature of a highly significant site while maintaining an international outlook.

The building was designed with local design firm Mitsubishi Jisho Sekkei as part of a wider redevelopment programme for the area. The Marunouchi district competes with two other principal business zones in Tokyo – Shiodome and Roppongi – so Mitsubishi Estates (the owner of the land) embarked on a project to convert this purely financial centre into a 'bustling twenty-four-hour district'. The completion of this tower was the culmination of the first phase of this redevelopment; the second began in 2008 with the modernization of three existing buildings.

Perfectly mirroring in silhouette the older Marunouchi Building across the road, the Shin-Marunouchi Tower, clad in dark aluminium grids, is much more streamlined (the 'podium' of the present Marunouchi Tower is modelled on the shape of the original, eight-storey building of 1923, designed by Kotaro Sakurai, who trained in the UK; the original was demolished in 1999 and the current form, with its recessed upper levels, emerged in 2002). Each square of the grid acts as a building block from which the new structure sprang, similar to the way in which Japanese houses were once modular, their dimensions based on *tatami* mats. The effect is to retain a human scale for the tower.

The mixed-use building, with about 195,000 square metres (2,100,000 sq. ft) of floor space, is constructed from a steel frame above ground, and reinforced concrete below. Its four subterranean levels are directly linked underground to the nearby railway station. A seven-storey podium containing leisure facilities and 150 stores, bars and restaurants supports a pair of towers, with offices up to the thirty-seventh floor. On the podium is an L-shaped oasis-like terrace, where office workers can enjoy meals high above the street.

Monolithic and muscular, the Shin-Marunouchi Tower has a powerful presence, although the stepped-back upper elements prevent it from becoming overly intimidating. It is also beautifully detailed, and its large atrium spaces are dramatic. The Marunouchi district accounts for about twenty per cent of Japan's GDP, so the tower had to reflect that economic status: as a new landmark, its office spaces pull in leading business tenants from around the world.

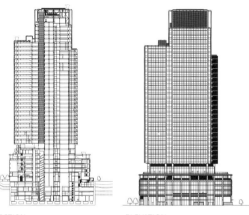

SECTION　　　　ELEVATION

OPPOSITE The tower in its business-district context. LEFT The building seen from outside the Imperial Palace's extended public garden. BELOW The tower sits on a plinth, its bulk emphasized by a horizontal recess. BOTTOM Atria are sensitively detailed and dramatic in scale.

OFFICES AND RETAIL 259

Showroom H
Joetsu, Niigata Prefecture

AKIHISA HIRATA ARCHITECTURE OFFICE, 2005–2006

An unusually elegant setting for the sale of agricultural machinery, Showroom H stands out against its backdrop of rice paddies and modest homes. It replaced a worn-out wooden building that housed the family-owned business. While the client, a forty-something architecture enthusiast, could have opted for a more conventional solution, he held an internet-based design competition instead. Architect Akihisa Hirata, who had just left Toyo Ito's office to launch his own practice, won the commission hands down.

What convinced the client of Hirata's suitability for the project was the photograph of fish swimming among coral tentacles that he used to explain his concept. Inspired by this image, Hirata's idea was to create a fluid space where, propelled by a sense of discovery, potential customers could move freely between partially hidden places, each one a display area for the latest in lawnmowers.

To convert this concept into concrete, Hirata created a simple rectangular box and inserted diagonal walls arranged in a 5-metre (16⅖-ft) grid, like a giant egg box. In lieu of full-height partitions that explicitly separate one bay from the next, he used triangular panels with tapered tips that barely touch the ground, simultaneously separating and connecting adjacent areas. At the perimeter, the angled planes pique the curiosity of passers-by, who can glimpse individual wares but must venture inside to view the full array – a clever marketing strategy Hirata picked up while working on retail design in Ito's office.

To incorporate the showroom's additional functions within the confines of the building's rigid geometry, Hirata lopped off a corner here and removed a partition there. Accessed from the main road in front, the principal entrance is recessed in a triangular bay excised from the box but protected by the flat roof. Filling the first floor are display areas, an office and a salon-style sales area, where customers can sip tea as they weigh the pros and cons of one snowblower over another. Open stairs in the middle of the building lead up to the smaller second floor, where Hirata located a reference room and a rest area for employees.

Bold and brawny, Hirata's unmitigated concrete provides the perfect contrast with the intricate machinery for sale inside. Unlike most showrooms, which are cluttered with signage and excess furniture, Hirata's minimalist space keeps the focus where it should be: on the goods. Its neutral grey surfaces elevate each piece of equipment into a thing of beauty.

OPPOSITE, TOP The showroom fronts a regional highway. OPPOSITE, BOTTOM A recessed entrance leads directly on to the showroom floor. RIGHT View out from the showroom. FAR RIGHT A wall-mounted scooter on display. BELOW Triangular wall panels both divide and connect the showroom's adjacent spaces.

Tod's Omotesando Building

Shibuya, Tokyo

TOYO ITO & ASSOCIATES, ARCHITECTS
2002–2004

The ritual of wrapping extends a long way in Japan, but can a shop be worn in the same way as a dress? In the past, Toyo Ito has spoken of his desire to create a permeable type of architecture that the users could (metaphorically speaking) wrap around themselves like a piece of clothing, constructing in the process a new identity: a new, modern self.

In 2002 the luxury Italian shoemaker Tod's commissioned Ito to build its flagship store on Omotesando's architectural catwalk in the heart of Tokyo. For Tod's, which is poised to seize a slice of the Japanese market, the architect's tight grasp of fashion and architecture and the regenerative abilities of both resulted in a winning scheme.

Ito's seven-storey building continues his explorations of surface and skin, and is immediately distinguished by its dramatic exterior form. The façade copies the shapes of the tall zelkova trees that line the street (Ito's second encounter with zelkova trees; see Sendai Mediatheque, pages 84–85) and reinterprets them as a series of criss-crossing geometric braces. At night, when the spaces between the braces are lit from within, the building appears to be gently wrapped in the silhouettes of the trees. During the day, however, the impression is quite different. The exterior façade is a smooth blend of concrete and glass. Some 450 tonnes of reinforcement bars were used, but they are so well concealed that the structure feels weightless.

The £70 million headquarters building offers impressive facilities for Tod's. The boutique sprawls over three floors; offices are on the two floors above. The fifth floor is dedicated to a single room for private events, and the top floor to a boardroom, which extends on to a terraced roof garden. Inside the boutique, the fixtures, fittings and furnishings – such as the small side tables where Tod's staff keep their shoehorns – flamboyantly echo the angular shapes of the openings on the façade. Otherwise, however, the store's interior space bears little relation to the exterior. Although some areas of the floor are recessed to create air pockets around the openings in the façade, each storey is still dominated by horizontal flooring, the rigidity of which is clearly visible through the complexity of the computer-simulated branches, which suggest a more irregular interior.

Despite its apparent minimalism, the building displays none of the solemnity and intellectual snobbery that typify much contemporary retail architecture. Here, there are simply humour, a lightness of touch and a great sense of the drama and fun of shopping.

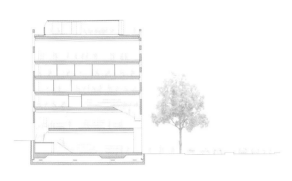

SECTION

PLANS: BASEMENT–SECOND FLOORS

PLANS: THIRD–SIXTH FLOORS

OPPOSITE The unusual bracing on the façade, echoing the branches of zelkova trees lining Omotesando Avenue, is load-bearing and enables column-free internal spaces. BELOW, LEFT Façade detail. BELOW, RIGHT The shop interiors look out on to Omotesando Avenue.

OFFICES AND RETAIL 263

Toririn

Yaizu, Shizuoka Prefecture

MOUNT FUJI ARCHITECTS STUDIO, 2004

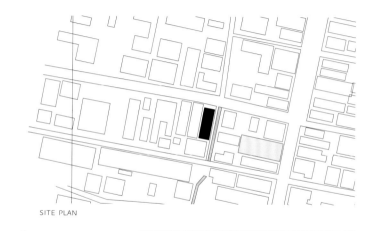

SITE PLAN

An industrial steel box with a twist, Toririn is the temporary home for a delicatessen and meat shop that has been in business for one hundred years. For much of its life, the store was a fixture on the town's main shopping street, but when government plans to widen the road were confirmed, the family-owned company had to move out and build a new shop.

Although the authorities contributed to the reconstruction, the only available plot near by was half the size and hemmed in by buildings and a polluted river. In addition, the façade of any building would look north. Faced with this predicament, the client turned to his former classmate Masahiro Harada in the hope that the Tokyo-based firm headed by Harada and his wife, Mao, would be able to work within these constraints. With such dark and dirty site conditions as these, the big question for the designers was how to edit the context but bring in daylight. Clearly the only option was from above.

Fresh from the success of their studio XXXX (pages 128–29), the architects again chose a repetitive-frame construction system, this time made of steel and torqued in the middle to open triangular slits on alternate sides of the roof instead of the walls. Clad inside and out with plasterboard, the four frames are made of welded steel pipe fixed to a concrete base. Thanks to the twist, the roof openings wash the shed-like interior with diffuse light all day, except at high noon, when direct rays fleetingly spotlight the shop's displays.

To maximize floor area and make room for scales and equipment of various vintages, the architects divided the 86-square-metre (926-sq.-ft) store as little as possible. While full-height glass doors open it completely to the street, a concrete-and-glass partition along the centreline – the only place where the ceiling is flat – separates the meat counters on one side from the delicatessen cases on the other. A walk-in refrigerator and small office fill out the back of the building.

In this incarnation, the shop may have only a short shelf life, since once the road construction finishes, in about five years' time, Toririn will return to new quarters on its former site. But when that day comes, this versatile building will most likely be reborn as a beauty parlour, cafe or other small-town commercial venture.

ABOVE The north-facing façade is entirely glazed, for visibility rather than for light. OPPOSITE, TOP LEFT While the side elevation is almost entirely devoid of openings, the sawtooth roof lets in light from above. OPPOSITE, TOP RIGHT The meat counter is bathed in indirect light. OPPOSITE, BOTTOM The site abuts a narrow stream.

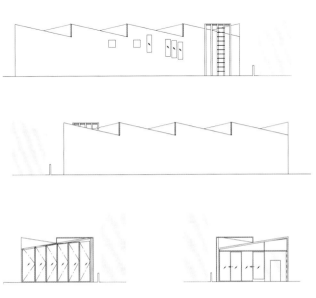

ELEVATIONS

OFFICES AND RETAIL 265

Triad

Hotaka, Nagano Prefecture

MAKI AND ASSOCIATES, 2000–2002

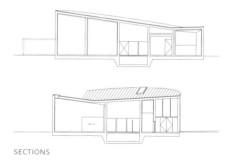

SITE PLAN

Triad, comprising three separate buildings with a combined floor space of 1100 square metres (11,840 sq. ft), is a relatively compact project in a range of large-scale works by Fumihiko Maki, one of the original members of the 1960s Metabolist group. These works include a university campus (which is more like a small town) in Singapore, Tower 4 of the World Trade Center complex in New York, and a cluster of buildings including a new station, a hotel and an office building in central Taipei, Taiwan.

Harmonic Drive, a company that produces high-tech precision parts for satellites, spacecraft and telescopes, commissioned Maki in 2000 to design a new laboratory, a private museum dedicated to the artist Yoshikuni Iida, and a security guard's office, all on a compound next to the company's main office. The setting could not be more idyllic: a Hokusai-like rural scene of farmers in straw hats toiling away in small fields appears against the backdrop of the snow-capped Japanese Alps.

Rejecting any nostalgia that might be inspired by such a landscape, the laboratory is crisply enclosed in a shell of reflective stainless steel. Its hyperbolic profile eliminates air pockets, ensuring an even temperature throughout. The deep-set canopy allows space over the main entrance for a small balcony, supported by a thin column painted in bright yellow, the only drop of colour in the whole compound. It perhaps signifies the start of sequential circular movements, which are repeated in the rounded wall that encloses the art gallery's reception area, as well as the footpaths that swerve around the three buildings.

The projectile mobility continues. Near the compound's main gate is the guardhouse, a simple rectilinear volume that cantilevers off the gentle slope. Much more compact than the other two buildings, it nevertheless plays a significant role in affecting the nearby gallery's composition and material palette. The sculpture gallery, located roughly halfway between the laboratory and the guardhouse, is neither completely quadrilateral nor cylindrical, but a hybrid. Meanwhile, the frosted glass panel that clads the cantilevered end of the guardhouse is repeated in a row covering half of the glazed façade under the gallery's pitched roof. Materials as well as dynamic forms intelligently bind the scheme's components together.

Maki's work has withstood the rapid growth and changes of our time, thanks to his modest attitude that individual buildings are just one small part of a collective whole. The architect is more like a masterful choreographer, creating ensemble pieces for dancers, who have an energy of their own.

GALLERY

SECTIONS

GUARDHOUSE

SECTIONS

OPPOSITE The sweeping volume of the laboratory eliminates air pockets. LEFT The guardhouse, a simple rectilinear volume, cantilevers off the gentle slope. BELOW, LEFT Triad is eclipsed by the magnificent Japanese Alps. BELOW, RIGHT A curvaceous footpath connects the gallery to the laboratory. BOTTOM LEFT The laboratory interior. BOTTOM RIGHT Triad maintains an exquisite tension between curves and straight lines.

LABORATORY

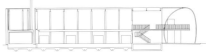

SECTIONS

OFFICES AND RETAIL 267

Further Reading

Books

Barthes, Roland, *Empire of Signs*, trans. Richard Howard, New York (Hill & Wang) 1983

Brayer, Marie-Ange, Migayrou, Frédéric, and Nanjo, Fumio, *ArchiLab's Urban Experiments: Radical Architecture, Art and the City*, London (Thames & Hudson) 2005

Buruma, Ian, *A Japanese Mirror: Heroes and Villains of Japanese Culture*, London (Penguin Books) 1988

Daniell, Thomas, *After the Crash: Architecture in Post-Bubble Japan*, New York (Princeton Architectural Press) 2008

Egashira, Shin, *Before Object, After Image: Koshirakura Landscape 1996–2006*, London (Architectural Association) 2006

Frampton, Kenneth, *Modern Architecture: A Critical History*, London (Thames & Hudson) 2007

Isozaki, Arata, *Japan-ness in Architecture*, ed. David B. Stewart, trans. Sabu Kohso, Cambridge, Mass. (MIT Press) 2006

Jansen, Marius B., *The Making of Modern Japan*, Cambridge, Mass. (Bellknap Press of Harvard University Press) 2000

Kaijima, Momoyo, Kuroda, Junzo, and Tsukamoto, Yoshiharu, *Made in Tokyo*, Tokyo (Kajima Institute Publishing Co.) 2001

Maki, Fumihiko, *Investigations in Collective Form*, St Louis (Washington University School of Architecture) 1964

Miyoshi, Masao, and Harootunian, H.D. (eds), *Postmodernism and Japan*, Durham, NC (Duke University Press) 1989

Pollock, Naomi, *Modern Japanese House*, London (Phaidon) 2005

Sand, Jordan, 'Street Observation Science and the Tokyo Economic Bubble, 1986–1999' in Gyan Prakash and Kevin Michael Kruse (eds), *The Spaces of the Modern City: Imaginaries, Politics, and Everyday Life*, Princeton, NJ (Princeton University Press) 2008

Sontag, Susan, *On Photography*, London (Penguin Classics) 2002

Sorensen, André, *The Making of Urban Japan: Cities and Planning from Edo to the Twenty-First Century*, London and New York (Routledge) 2002

Stewart, David B., *The Making of a Modern Japanese Architecture: From the Founders to Shinohara and Isozaki*, Tokyo and New York (Kodansha International) 2002

Suzuki, Hiroyuki, Banham, Reyner, and Kobayashi, Katsuhiro, *Contemporary Architecture of Japan 1958–1984*, New York (Rizzoli) 1985

Tanizaki, Junichiro, *In Praise of Shadows*, New Haven, Conn. (Leete's Island Books) 1977

Tsuzuki, Kyoichi, *Tokyo: A Certain Style*, San Francisco, Calif. (Chronicle Books) 1999

Articles

Kaji-O'Grady, Sandra, 'Authentic Japanese Architecture after Bruno Taut: The Problem of Eclecticism', *Fabrications* 11, no. 2, 2001

Oe, Kenzaburo, 'Japan, the Ambiguous, and Myself', Nobel Lecture, 7 December 1994, nobelprize.org/nobel_prizes/literature/laureates/1994/oe-lecture.html

Williamson, Alan, 'Koan Ranger', interview with Gary Snyder, Poetry Foundation, 29 May 2009, poetryfoundation.org/journal/article.html?id=181667

Yatsuka, Hajime, 'Autobiography of a Patricide: Arata Isozaki's Initiation into Postmodernism', in Thomas Weaver (ed.), *AA Files 58*, 2009

Index

Page numbers in **bold** refer to main project entries.

21_21 Design Sight *see* Tadao Ando Architect & Associates
21st Century Museum of Contemporary Art, Kanazawa *see* Kazuyo Sejima + Ryue Nishizawa/SANAA

AAT and Makoto Yokomizo, Architects
  Tomihiro Art Museum 14–15, **88–89**
Abe, Hitoshi 27, 74, 110, 234, **238–39**
Aichi Expo 2005: Global Loop (Kikutake) 11, 13
  Spanish Pavilion **118–19**
Aichi Prefecture 12, 30–31, 120, 121, 158–59, 228–29
Akamatsu, Kazuko 146–47
Akao, Kazuo 204
Akihisa Hirata Architecture Office:
  Gallery Sora **68–69**
  Showroom H 24, **260–61**
Aláez, Ana Laura 91
Alphaville: Hall House I **194–95**
Amorphe: Chapel Aktis **174–75**
  Rikuryo Alumni Hall, Kitano High School **162–63**
Ando, Tadao 12, 14, 26, 27, **52–53**, **58–59**, **64–65**, **140–41**, 242, **244–45**, 246
Ando, Yuko 182, 212
Antwerp, Belgium: deSingel art gallery 48
Aoba-Tei *see* Atelier Hitoshi Abe
Aoki, Jun 18, **62–63**, **114–15**, **166–67**, 174, 182, **190–91**, **212–13**, 250
Aomori-ken Dog (Nara) 62
Aomori Museum of Art *see* Jun Aoki & Associates
Aomori Prefecture 12, 18, 62–63, 90–91
Archigram 11
Architects Teehouse 84, 85
Architectural Association School of Architecture, London 12
  Koshirakura Workshop Projects **76–77**
Armani, Giorgio 244

Artpolis 25
Arup Japan: Inujima Art Project: Seirensho **70–73**
Atelier BNK: Sapporo Dome **116–17**
Atelier Bow-Wow 7, 16–17, 80, 121
  Gae House 6, 16, **192–93**
  Hanamidori Cultural Centre 16–17, 35, **38–39**
  House Crane **196–99**
  House and Office 23, 24, **248–49**
Atelier Hitoshi Abe: Aoba-Tei **234–35**
  F-Town **238–39**
  Kanno Museum of Art **74–75**
  Miyagi Stadium **110–11**
Atelier φ: Sapporo Dome **116–17**
Atsushi Kitagawara Architects 121
  Gifu Academy of Forest Science and Culture, Mino **144–45**
aTTa (Tsubaki) 91

Bacon, Francis 78
Baisouin Temple *see* Kengo Kuma & Associates
Barthes, Roland 11
Bellini, Mario 112
Benesse Corporation 70
Berlin: Internationale Bauausstellung 25
Big Window House *see* Tezuka Architects
Bilbao, Spain: Guggenheim Bilbao 54
Blackboard Classroom (Kawaguchi) 81
Blue Planet Sky (Turrell) 60
Bridge of Light (Aláez) 91
Brillare Dining and Party Room *see* Klein Dytham Architecture
Bubbletecture M *see* Endo Shuhei Architect Institute
Buckminster Fuller, Richard 11, 134
Buddhist Temple *see* Heatherwick Studio

C + A (Coelacanth and Associates) 146
Calvin Klein 226
CAn + CAt: Gunma Kokusai Academy **146–47**
Canopy House *see* Tezuka Architects

Centrair *see* Nikken Sekkei
Chagall, Marc 62
Chapel Aktis *see* Amorphe
Chiba Prefecture **104–105**
Chichu Art Museum *see* Tadao Ando Architect & Associates
Children's Centre for Psychiatric Rehabilitation *see* Sou Fujimoto Architects
Choi, Jeongha 91
Chokkura Plaza *see* Kengo Kuma & Associates
City in the Air *see* Arata Isozaki

Daigo Ishii + Future-scape Architects House of Light **104–105**
Daniell, Thomas 13–14
De Maria, Walter 64
Dior, Christian 236
Dior Omotesando Store *see* Kazuyo Sejima + Ryue Nishizawa/SANAA
Dublin City Art Gallery The Hugh Lane 78
Dytham, Mark 178

Earth Summit scheme (Rio de Janeiro, Brazil, 1992) 11
East Japan Railway Company 40
Echigo-Tsumari Art Triennial 80
Echigo-Matsunoyama Museum of Natural Science *see* Tezuka Architects
EDH (Endoh Design House)
  Natural Strips II 27, **214–15**
Egashira, Shin 12
  Koshirakura Workshop Projects **76–77**
Endo, Shuhei 12, **36–37**, **50–51**, **134–35**, **220–21**
Endo Shuhei Architect Institute:
  Bubbletecture M **134–35**
  Halftecture OJ/OO/OR **36–37**
  Rooftecture S 12, **220–21**
  Slowtecture M **50–51**
Endoh, Masaki 27, 214
Engawa House *see* Tezuka Architects
Erlich, Leandro 60, 61

Final Wooden House *see* Sou Fujimoto Architects
Flower Horse (Choi) 91
Flower Shop H *see* Office of Kumiko Inui
Foreign Office Architects: Spanish Pavilion/Aichi Expo 2005 **118–21**
  Yokohama International Port Terminal 25, 28–29, **54–55**
Foster + Partners: Kamakura House **204–207**
Fragments of Color Cubes (Takahashi) 91
Frampton, Kenneth 11
F-Town *see* Atelier Hitoshi Abe
Fuji, Atsushi 102
Fuji, Jeanie 102
Fuji Kindergarten *see* Tezuka Architects
Fujie, Kazuko 164
Fujimori, Terunobu 7, 15
  Lamune Onsen **108–109**
  Nemunoki Museum of Art **82–83**
  Takasugi-an 10, 16, **94–95**, **122–23**
  Yakisugi House **230–31**
Fujimoto, Sou 7, 12, 17, 25, 34, **100–101**, **160–61**, **176–77**, **184–85**, **200–201**, **202–203**
Fujiya Inn *see* Kengo Kuma & Associates
Fukuoka Prefecture 24, 26, 34–35, 38
Fukusaki Hanging Garden *see* Kengo Kuma & Associates
Fukutake, Soichiro 70, 71, 246
Fukutake Hall, Tokyo University, Hongo Campus *see* Tadao Ando Architect & Associates
Future University – Hakodate *see* Riken Yamamoto & Field Shop

G *see* Jun Aoki & Associates
Gae House *see* Atelier Bow-Wow
Gallery Sora *see* Akihisa Hirata Architecture Office
Gaudí, Antonio 250
Gifu Prefecture 16, 42–45, 144–45
Gilbert & George 204
Gluckman Mayner Architects 27

269

Golgi Structures (Maki) 11
Gravitecture series (Endo) 36
Great Hanshin Earthquake (1995) 50
Grin Grin see Toyo Ito & Associates, Architects
Gunma Prefecture 14–15, 88–89, 146–47
Gyre see MVRDV

Halftecture OJ/OO/OR see Endo Shuhei Architect Institute
Hall House I see Alphaville
Hanamidori Cultural Centre see Atelier Bow-Wow
Hara, Hiroshi 46
  Sapporo Dome **116–17**
Harada, Mao 264
Harada, Masahiro 264
Heatherwick, Thomas **172–73**
Heatherwick Studio: Buddhist Temple **172–73**
Helix City (Kurakawa) 11
Hermès 250
Herzog, Jacques 256
Herzog & de Meuron: Prada Aoyama 17, **256–57**
hh Style see Tadao Ando Architect & Associates
Hirata, Akihisa 24, **68–69**, **260–61**
Hiroshima 2045: City of Peace and Creativity 25, 46
Hiroshima City 12, 15, 18, 25, 26, 46–47
HOK: Centrair **30–31**
Hokkaido 17, 25, 26, 106–107, 116–17, 124–27, 142–43, 176–77, 216–17
Honmura Lounge and Archive see Office of Ryue Nishizawa
Honshu 62
Hopkins Architects: Shin-Marunouchi Tower **258–59**
Horiguchi, Sutemi 17
Hoshakuji Station see Kengo Kuma & Associates
Hoshino, Tomihiro 19–22, 88
Hosokawa, Morihiro 25
House Crane see Atelier Bow-Wow
House of Light see James Turrell
House N see Sou Fujimoto Architects
House O see Sou Fujimoto Architects
House and Office for Atelier Bow-Wow see Atelier Bow-Wow
Hyogo Prefecture 12, 50–51, 220–21

Igarashi, Jun 26, **124–27**, **216–17**
Ikeda, Masahiro: Natural Strips II 27, **214–15**
Ina-Higashi Elementary School see Mikan
Inui, Kumiko 7, 15, **240–41**
Inujima Art Project: Seirensho see Sambuichi Architects
Irie, Kei'ichi 12, **228–29**

Ishigami, Junya 7, 15, **150–51**
Ishikawa Prefecture 14, **60–61**, 88, 91
Isozaki, Arata 13, 17–18, 25, 32, 84
  City in the Air 11
Ito, Toyo 14, 16, 18, 24, **34–35**, **38–39**, **42–45**, 62, **84–85**, 88, 100, **130–31**, **164–65**, 234, 236, **250–51**, 260, **262–63**

Jansen, Marius B. 11
Japanese Alps 266, 267
Jigsaw 226
Jinchokan Moriya Historical Museum 122
Jun Aoki & Associates: Aomori Museum of Art 18, **62–63**
  G **190–91**
  N **212–13**
  OPJ **114–15**
  White Chapel **182–83**
Jun Igarashi Architects: Rectangle of Light **216–17**
  Ware House 25, 26, **124–27**
Junya Ishigami + Associates: Kaito Workshop 15, **150–51**

Kagawa Prefecture 20–21, 48–49, 64–65, 150, 246–47
Kagoshima Prefecture 172–73
Kaijima, Momoyo 16, 38
Kaito Workshop see Junya Ishigami + Associates
Kamakura House see Foster + Partners
Kanagawa Prefecture 15, 25, 28–29, 54–55, 92–93, 150–51, 186–87, 204–207, 208–209, 212–13
Kanno Museum of Art see Atelier Hitoshi Abe
Kawabata, Yasunari 13
Kawaguchi, Tatsuo 81
Kazuyo Sejima + Ryue Nishizawa/SANAA: 21st Century Museum of Contemporary Art, Kanazawa 14, **60–61**, 88, 91
  Naoshima Ferry Terminal **48–49**, 150, 246
Kei'ichi Irie + Power Unit Studio: Y House 12, **228–29**
Kenchiku Tanteidan (Architecture Detectives) 16
Kengo Kuma & Associates: Baisouin Temple **168–71**
  Chokkura Plaza **32–33**
  Fujiya Inn **102–103**, 107
  Fukusaki Hanging Garden **138–39**
  Hoshakuji Station 32, **40–41**
  Lotus House **208–209**
  Masanari Murai Memorial Museum of Art **78–79**
  One Omotesando **252–55**
  Steel House **224–25**
  Suntory Museum of Art 27, **86–87**

Kikutake, Kiyonori 11, 13
Kitagawara, Atsushi **144–45**
Klein, Astrid 178
Klein Dytham Architecture 27
  Brillare Dining and Party Room **96–99**, 112, 178
  Leaf Chapel 96, 112, 174, **178–79**
  Moku Moku Yu **112–13**, 178
Koga Park Cafe (SANAA) 150
Kohn Pedersen Fox Associates 27, 168
Kojima, Kazuhiro 146–47
Koshirakura Workshop Projects see Shin Egashira
Kuma, Kengo 18, 27, **32–33**, **40–41**, **78–79**, **86–87**, **102–103**, 107, **138–39**, **168–71**, **208–209**, **224–25**, **252–55**
Kumamoto Prefecture 8–9, 17, 25, 100–101
Kumamura Forestry Association 100
Küng, Moritz 48
Kurakawa, Kisho 11, 12, 168
Kuramata, Shiro 226
Kuramure see Makoto Nakayama
Kuroda, Junzo 38
Kyoto Prefecture 15, 26, 140, 152–53, 172, 174–75
Kyushu 17, 100, 200

Lambie, Jim 91
Lamune Onsen see Terunobu Fujimori
Le Corbusier 125
Leaf Chapel see Klein Dytham Architecture
Lin, Michael 91
London: British Museum 54
Lotus House see Kengo Kuma & Associates
Lovegrove, Ross 85

Maki, Fumihiko 11, 13, 242, **266–67**
Maki and Associates 242
  Triad **266–67**
Masanari Murai Memorial Museum of Art see Kengo Kuma & Associates
Matsudai Snow-Land Agrarian Culture Centre see MVRDV
Meiso no Mori Municipal Funeral Hall see Toyo Ito & Associates
Metabolists 11, 266
MIAS: Engawa House **188–89**
Mies van der Rohe, Ludwig 250
Mikan: Ina-Higashi Elementary School **132–33**, **148–49**
Mikimoto, Kokichi 250
MIKIMOTO Ginza 2 see Toyo Ito & Associates, Architects
Minami-Yamashiro Primary School see Richard Rogers Partnership
Ministry of Economy, Trade and Industry 70
Mishima, Yukio 12, 71
Mitsubishi Jisho Sekkei 258

Miyagi, Mariko 83
Miyagi Prefecture 14, 26, 38, 74–75, 84–85, 110–11, 180–81, 234–35, 236, 238–39, 262
Miyagi Stadium see Atelier Hitoshi Abe
Miyajima, Tatsuo 48
Miyake, Issey 58
Mode Gakuen Cocoon Tower see Tange Associates
Mode Gakuen Spiral Towers see Nikken Sekkei
Moku Moku Yu see Klein Dytham Architecture
Monet, Claude 64
Moriyama House see Office of Ryue Nishizawa
Morrison, Paul 91
Mount Asama 218
Mount Fuji Architects Studio: Toririn **264–65**
  XXXX **128–29**, 264
Mueck, Ron 56–57, 91
Murai, Masanari 78, 79
Musashino Art University Library see Sou Fujimoto Architects
MVRDV: Gyre **242–43**, 244
  Matsudai Snow-Land Agrarian Culture Centre **80–81**

N see Jun Aoki & Associates
Nabeshima, Chie 218
Nagano Prefecture 10, 15–16, 32–33, 94–95, 122–23, 148–49, 196–99, 218–19, 222–23, 230–31, 266–67
Nagoya Isen 158
Nagoya Mode Gakuen 158
Naka Incineration Plant see Taniguchi and Associates
Nakata, Hideo 102
Nakayama, Makoto: Kuramure **106–107**
Naoshima Ferry Terminal see Kazuyo Sejima + Ryue Nishizawa/SANAA
Naoshima Fukutake Art Museum Foundation 48, 64, 246
Nara, Yoshitomo 62
Natural Strips II see EDH
Nemunoki Museum of Art see Terunobu Fujimori
New York: MOMA extension 46
  New Museum 237
  Tower 4, World Trade Center complex 266
Niigata Prefecture 12, 24, 66–67, 76–77, 80–81, 104–105, 260–61
Nikken Sekkei: Centrair **30–31**
  Mode Gakuen Spiral Towers **158–59**
  Tokyu-Toyoko Line Shibuya Station **52–53**, 140, 214
Nishizawa, Ryue 7, 12, 14, **48–49**, **60–61**, **90–91**, **210–11**, **236–37**, **246–47**
Nuno Corporation 182, 212

Obunsha Group 204
Ocean City (Kikutake) 11
Ochrea (Morrison) 91
Oe, Kenzaburo 13
Office of Kumiko Inui: Flower Shop H **240–41**
Office of Ryue Nishizawa: Honmura Lounge and Archive/Benesse Art Site Naoshima **246–47**
   Moriyama House 14, **210–11**
   Towada Art Center 12, **90–91**
Oita Prefecture 108–109, 200–201
Okayama Prefecture 70–73, 230
Okinawa Prefecture 114–15
One Omotesando *see* Kengo Kuma & Associates
Ono, Yoko 91
OPJ *see* Jun Aoki & Associates
Osaka Prefecture 12, 26, 36–37, 52, 118, 138–39, 162–63, 166–67, 182

Pawson, John: Tetsuka House **226–27**
Perry, Matthew 92
Piano, Renzo 250
Power Unit Studio 12, **228–29**
Prada Aoyama *see* Herzog & de Meuron

Rashid, Karim 85
Rectangle of Light *see* Jun Igarashi Architects
Richard Rogers Partnership: Minami-Yamashiro Primary School **152–53**
Riken Yamamoto & Field Shop: Future University – Hakodate **142–43**
   Yokosuka Museum of Art **92–93**
Rikuryo Alumni Hall, Kitano High School *see* Amorphe
Ring House *see* TNA
Rogers, Richard **152–53**
Rogers Stirk Harbour + Partners: Minami-Yamashiro Primary School **152–53**
ROJO (Roadway Observation Society) 16
Roof House *see* Tezuka Architects
Rooftecture S *see* Endo Shuhei Architect Institute
Rossi, Aldo 24
Royal Institute of British Architects 152

Sakurai, Kotaro: Marunouchi district, Tokyo 258
Sambuichi, Hiroshi 7, 12, 13, **70–73**
Sambuichi Architects: Inujima Art Project: Seirensho **70–73**
SANAA 13, 18, 19–21, 84, 90, 210
   21st Century Museum of Contemporary Art, Kanazawa 14, **60–61**, 88, 91
   Dior Omotesando Store **236–37**, 244
   Naoshima Ferry Terminal **48–49**, 150, 246
Sannai Maruyama archaeological site 18, 62

Sapporo Dome *see* Hiroshi Hara
Sasaki, Mutsuro 16, 34–35, 42, 84, 250
Sejima, Kazuyo **48–49**, **60–61**, 85, 88, 91, 150, **236–37**
Sendai Baptist Church *see* Soy Source Architects
Sendai Mediatheque *see* Toyo Ito & Associates
Seto 118
Shiga Prefecture 134–35, 194–95
Shimazaki, Takero 7
Shin-Marunouchi Tower *see* Hopkins Architects
Shizuoka Prefecture 82–83, 128–29, 264–65
Showroom H *see* Akihisa Hirata Architecture Office
Singapore 266
Skidmore, Owings & Merrill 27
Slowtecture M *see* Endo Shuhei Architect Institute
Snyder, Gary 15
Sontag, Susan 18
Sou Fujimoto Architects: Children's Centre for Psychiatric Rehabilitation 17, 25, **176–77**
   Final Wooden House 8–9, 17, **100–101**
   House N **200–201**
   House O **202–203**
   Musashino Art University Library **160–61**
Soy Source Architects: Sendai Baptist Church **181–81**
Spanish Pavilion/Aichi Expo 2005 *see* Foreign Office Architects
Stage House *see* TNA
Standing Woman (Mueck) 56–57, 91
Starck, Philippe 24
Steel House *see* Kengo Kuma & Associates
Sugimoto, Hiroshi 48
Suh, Do-Ho 91
Suntory Museum of Art *see* Kengo Kuma & Associates
Super Car School 16–17
Super-OS: Matsudai Snow-Land Agrarian Culture Centre **80–81**
The Swimming Pool (Erlich) 60, 61

Tadao Ando Architect & Associates:
   21_21 Design Sight 26, 27, **58–59**
   Chichu Art Museum 20–21, **64–65**, 246
   Fukutake Hall, Tokyo University, Hongo Campus **140–41**
   hh Style **244–45**
   Tokyu-Toyoko Line Shibuya Station **52–53**, 140, 214
Taipei, Taiwan 266
Taisei Corporation 116
Takahashi, Kyota 91
Takamori, Saigo 172
Takasugi-an *see* Terunobu Fujimori

Takei, Makoto 218
Takenaka Corporation 116
Tama Art University Library *see* Toyo Ito & Associates, Architects
Takeyama, Kiyoshi Sey 162, 174
Tange, Kenzo 11, 18, **154–57**
Tange Associates: Mode Gakuen Cocoon Tower **154–57**, 158
Taniguchi, Yoshio 13, 15, 25, **46–47**
Taniguchi and Associates: Naka Incineration Plant 15, 25, **46–47**
Tanizaki, Junichiro 106
Tetsuka House *see* John Pawson
Tezuka, Takaharu 137
Tezuka Architects 7, 218
   Big Window House 15, **186–87**
   Canopy House 186
   Echigo-Matsunoyama Museum of Natural Science **66–67**
   Engawa House **188–89**
   Fuji Kindergarten **136–37**
   Roof House 186
Time/Timeless/No Time (De Maria) 64
TNA 15, 222–23
   Ring House **218–19**
   Stage House **222–23**
Tochigi Prefecture 32–33, 40–41
Tod's Omotesando Building *see* Toyo Ito & Associates, Architects
Tokoto, Asao 238
Tokyo 6, 11, 12, 14, 16, 17, 22, 23, 24, 26–27, 32, 35, 38–39, 52–53, 58–59, 68–69, 78–79, 80, 86–87, 92, 130–31, 136–37, 140–41, 142, 154–57, 158, 160–61, 164–65, 168–71, 188–89, 190–91, 192–93, 204, 210–11, 214–15, 224–25, 226–27, 232–33, 236–37, 240–41, 242–43, 244–45, 248–49, 250–51, 252–55, 256–57, 258–59, 262–63
Tokyu Architects and Engineers 52–53
Tokyu Corporation 52–53
Tokyu-Toyoko Line Shibuya Station *see* Tadao Ando Architect & Associates
Tomihiro Art Museum *see* AAT and Makoto Yokomizo, Architects
Toririn *see* Mount Fuji Architects Studio
Toshogu Shrine 15
Towada Art Center *see* Office of Ryue Nishizawa
Toyo Ito & Associates: Grin Grin **34–35**, 38
   Hanamidori Cultural Centre 16, 35, **38–39**
   Meiso no Mori Municipal Funeral Hall 16, **42–45**
   MIKIMOTO Ginza 2 **250–51**
   Sendai Mediatheque 14, 38, **84–85**, 234, 236, 262
   Tama Art University Library **164–65**
   Tod's Omotesando Building 22, 24, 232–33, 236, **262–63**
   Za-Koenji Public Theatre **130–31**

Triad *see* Maki and Associates
Tsubaki, Noboru 91
Tsukamoto, Yoshiharu 16, 38
Tsukihashi, Osamu 85
Tsulamoto Laboratory, Tokyo Institute of Technology
   Gae House 6, 16, 192, **192–93**
Turrell, James 60, 64
   House of Light **104–105**

University of California Department of Architecture and Urban Design 27
Uno, Susumu 146–47

Vuitton, Louis 250, 252

Wall-less House (Tezuka Architects) 186
Ware House *see* Jun Igarashi Architects
Webb, Michael 84
White Chapel *see* Jun Aoki & Associates
Wish Tree for Towada (Ono) 91
Wright, Frank Lloyd 32

XXXX *see* Mount Fuji Architects Studio

Y House *see* Kei'ichi Irie + Power Unit Studio
Yakisugi House *see* Terunobu Fujimori
Yamagata Prefecture **102–103**, 107
Yamamoto, Riken 46
Yamanashi Prefecture 96–99, 112–13, 174, 178
Yanagi, Yukinori 12, 70
Yatsuka, Hajime 15
Yokohama International Port Terminal *see* Foreign Office Architects
Yokomizo, Makoto 14–15, 88
Yokosuka Museum of Art *see* Riken Yamamoto & Field Shop

Zaera-Polo, Alejandro 118
Za-Koenji Public Theatre *see* Toyo Ito & Associates, Architects

Acknowledgements

This book showcases roughly one hundred buildings that Edmund and I have visited in Japan over several years. It would not have been possible to take on such an ambitious task without the help of many people who readily offered advice and expertise. In particular, we should like to thank Hugh Merrell and Julian Honer for commissioning the book; also at Merrell Publishers, Claire Chandler, Rosanna Lewis, Nicola Bailey and Nick Wheldon; Naomi Pollock and David Littlefield; Takero Shimazaki of Toh Shimazaki Architects for writing the foreword; Yvonne Jordan and Sophia Gibb of VIEW for organizing and sending images so swiftly; the Daiwa Anglo-Japanese Foundation and the Great Britain Sasakawa Foundation for their generous grants; All Nippon Airways for substantial discounts on flights to Japan; Leaf Digital for providing top-of-the-range camera equipment; and all the participating architects, without whose support and coorporation this book could not exist in its comprehensive form.

We should also like to extend our thanks to Professor Marie Conte-Helm, Director General of the Daiwa Anglo-Japanese Foundation, and Tomoko Kawamura, House Programme Director at the same organization, for the opportunity to curate and exhibit our work from Japan at Cornwall Terrace; Takao Anzawa, Cultural Attaché at the Embassy of Japan, and Simon Wright, Senior Coordinator for Cultural Affairs at the same institution, for their encouragement; Professor Adrian Forty at the Bartlett, University College London, for his luminous seminars on the various theories at work behind the creation of architectural history; Thomas Weaver, editor of the *AA Files* at the Architectural Association, for suggesting books, and for his thoughtful reading of my introductory text; Paul Stelmaszczyk, Head of Communications at Rogers Stirk Harbour + Partners, Nora Yamada, Jenny Turner and Alice Malik, for their direct editorial input; Shuhei Endo of Endo Shuhei Architect Institute for providing a list of architects working in the Kansai area of Japan; Naoko Kawamura of Kawamura Office, Noriko Tsukui, Senior Editor of *a+u* magazine, architectural journalist Taro Igarashi, and Osamu Tsukihashi of Teehouse Architects, editor of *Kenchiku Note* magazine, for participating in an informative discussion on the contemporary Japanese architectural scene over a delicious dinner in Tokyo; Akira Koyama of Key Operation for introducing Yoshiharu Tsukamoto of Atelier Bow-Wow to us years ago, kick-starting our career in Japan; Indy Priestman, Mizue Yoshimura and Jo McNeill, each of whom played a key role in enabling us to progress with our project; and finally my dear parents, Tsutomu and Hiroko, who allowed us to use their house in Tokyo as our HQ and to add a few digits to the odometer of their Toyota.

We dedicate this book to our children – Cosmo and Lulu – and in loving memory of our first daughter, Sakura.

Yuki Sumner
Edmund Sumner

London, September 2009

Biographies

Edmund Sumner is a London-based photographer, widely recognized as one of the best in his field. His work appears in publications all around the world. In 2003 he was interviewed for a documentary on CNN's Design 360. He also pursues and exhibits his own personal line of work. His most recent exhibitions include *Human Landscape* at Browse and Darby, and *Outside the Box* at the Daiwa Anglo-Japanese Foundation, both in London.

Yuki Sumner is a writer on contemporary art and architecture. Since 2002 she has worked with Edmund on projects in Japan. Her work is published regularly in newspapers and periodicals, including *Architectural Review*, *Wallpaper**, *ICON*, *The Guardian* and the *Sunday Telegraph*. She is currently completing her MA in architectural history at the Bartlett, University College London.

Naomi Pollock is an American architect based in Tokyo, and writes extensively on Japanese architecture. Her articles have appeared in numerous publications, including the *Financial Times*, the *New York Times*, *Wallpaper** and *Architectural Record*, for which she is Special International Correspondent. She is the author of *Modern Japanese House* (2005) and *Hitoshi Abe* (2009).

David Littlefield is an architectural writer based in England. He is the co-author of *Architectural Voices: Listening to Old Buildings* (2007) and the author of *Liverpool One: Remaking a City Centre* (2009). He is a visiting lecturer at the University of the West of England, and has also taught at the University of Bath and the University of the Arts London.

Takero Shimazaki studied architecture at the University of Wales, Cardiff, and the Bartlett, University College London. He worked for Itsuko Hasegawa Atelier and Richard Rogers Partnership before co-founding London-based practice t-sa (Toh Shimazaki Architects) with Yuli Toh. He is currently a visiting lecturer at Oxford Brookes University's Department of Architecture, the University of East London and TU Graz, Austria.